About Island Press

Island Press, a nonprofit organization, publishes, markets, and distributes the most advanced thinking on the conservation of our natural resources—books about soil, land, water, forests, wildlife, and hazardous and toxic wastes. These books are practical tools used by public officials, business and industry leaders, natural resource managers, and concerned citizens working to solve both local and global resource problems.

Founded in 1978, Island Press reorganized in 1984 to meet the increasing demand for substantive books on all resource-related issues. Island Press publishes and distributes under its own imprint and offers these services to other nonprofit organizations.

Support for Island Press is provided by Apple Computers Inc., Mary Reynolds Babcock Foundation, Geraldine R. Dodge Foundation, The Charles Engelhard Foundation, Ford Foundation, Glen Eagles Foundation, The George Gund Foundation, William and Flora Hewlett Foundation, The Joyce Foundation, The John D. and Catherine T. MacArthur Foundation, The Andrew W. Mellon Foundation, The Joyce Mertz-Gilmore Foundation, The New-Land Foundation, The J. N. Pew, Jr. Charitable Trust, Alida Rockefeller, The Rockefeller Brothers Fund, The Florence and John Schumann Foundation, The Tides Foundation, and individual donors.

WILDLIFE
AND HABITATS
IN MANAGED
LANDSCAPES

ED HOAG

EDITED BY JON E. RODIEK
AND ERIC G. BOLEN

Foreword by Laurence R. Jahn, President,
Wildlife Management Institute

WILDLIFE
AND HABITATS
IN MANAGED
LANDSCAPES

ISLAND PRESS

Washington, D.C. □ *Covelo, California*

The author is grateful for permission to include the following material, which appeared in *Transactions of the Fifty-fourth North American Wildlife and Natural Resources Conference,* copyright 1989: Chapter 4, Browse Diversity and Physiological Status of White-tailed Deer During Winter; Chapter 5, Conservation of Rain Forests in Southeast Alaska: Report of a Working Group; Chapter 6, American Martin: A Case for Landscape-level Management; and Chapter 7, Planning for Basin-level Cumulative Effects in the Appalachian Coal Field.

Reprinted by permission from the Wildlife Management Institute

Library of Congress Cataloging-in-Publication Data

Wildlife and habitats in managed landscapes : an overview / edited by
  Jon E. Rodiek and Eric G. Bolen : foreword by Laurence R. Jahn
     p.   cm.
  Includes index.
  ISBN 1-55963-053-1 (alk. paper). — ISBN 1-55963-052-3 (pbk. :
alk. paper)
  1. Wildlife habitat improvement—United States.   2. Habitat
(Ecology)—United States.   3. Natural resources—United States—
Management.   4. Landscape ecology—United States.   5. Land use,
Rural—Environmental aspects—United States.   I. Rodiek, Jon E.
QL84.2.W54   1991
639.9'2'0973—dc20   90-41593
                                                                    CIP

Printed on recycled, acid-free paper

Manufactured in the United States of America

10   9   8   7   6   5   4   3   2   1

This book is dedicated to Mr. Edmund C. and
Mrs. Elizabeth Books Rodiek
and to the memory of
Mr. and Mrs. W. F. Bolen

Acknowledgments

In addition to the authors, we would like to acknowledge the contributions made by the following people:

LAURENCE JAHN, Wildlife Management Institute

RICHARD E. McCABE, Wildlife Management Institute

LONNIE L. WILLIAMSON, Wildlife Management Institute

JO ANN GREGG, Texas A&M University

ED HOAG, Texas A&M University

MICHAEL McCARTHY, Texas A&M University

ELIZABETH D. BOLEN, Wilmington, NC

J.E.R.
E.G.B.

Contents

Foreword

A S HUMAN populations and technology continue to expand and dominate the landscape, more citizens express their concerns and demands for habitats that perpetuate wild living resources and associated outdoor experiences.

Prior to enactment of the National Environmental Policy Act in 1969, plans for developing and using lands, waters, and other natural resources usually were oriented toward a single purpose and focused on one scale—the immediate site. Insights gained in the next two decades broadened the horizons for planning; these lessons now emphasize the need for involving multiple scales at the earliest stages of comprehensive rather than single-sector planning. As a result, increasing numbers of citizens, professionals, and decision-makers now seek a better understanding of those concepts and procedures that will prevent environmental degradation and will achieve sustainable use with integrated resource management.

Although the concept of integrated resource management has existed for a century or more, that idea was seldom practiced. Instead, the resource base too often was exploited to meet current, short-term goals. However, as demands for more resources continued to expand, taxpayers were faced with mounting costs for restoring damaged areas, frequently including those developed using single-purpose plans. Chapters in this timely volume thus emphasize the broad search for better-designed resource developments, uses, and management.

Contributors to this book call for a realignment of planning procedures, thereby embracing multiple scales and a broad spectrum of societal values, needs, and interests. Sustaining actions are sought—a further weaning from merely exploitive uses of the resource base. Hence, a broadened awareness now calls for landscape plans, designs, and management procedures that will perpetuate ecological processes, conserve fish and wildlife populations, and protect other desirable features of the natural landscape. This concept for planning and management is advanced as "the ecosystem approach" and has been championed widely at sites such as the Great Lakes and the area encompassing Yellowstone National Park (e.g., Greater Yellowstone Ecosystem).

Following decades (1909 through 1978) of single-sector management, the Great Lakes Water Quality Board in 1978 recommended the ecosystem approach for collecting and using data. The board believed this approach to management was required to correct an archaic mind-set. Historically, water quality and its related objectives were considered solely in terms of water chemicals, largely one by one—an outmoded perspective that changed with adoption of the ecosystem approach. Water quality now is regarded as dynamic and the result of combined loading characteristics, which together form societal as well as biological consequences for the Great Lakes ecosystem. With its formation in 1978, the Great Lakes Water Agreement became one of the first international efforts to adopt the concept of ecosystem management.

The ecosystem approach acknowledges human interactions with, and impacts on, the resource base, rather than viewing people and their activities as separate from natural ecological processes. Participating governments now consider interactions of the Great Lakes with (a) areas within and neighboring the basin, (b) geochemical cycles and food chains, human needs, activities, interests, attitudes, perceptions, and desires, and (c) the carrying capacity, resilience, and recovery potential of water, land, air, and wild-living resources. Nonetheless, the search continues for measures of water quality that more adequately describe the well-being of the Great Lakes ecosystem. Integrative parameters, such as the health of fish stocks, hold much promise. In sum, introduction of the ecosystem concept represented a substantial breakthrough in managing the resources of the Great Lakes, but more work remains before the new management effort becomes fully operational.

Aligning human activities compatibly with landscape features, ecological processes, and wildlife habitat is a pressing challenge, as the authors of this volume emphasize. That it can be done is illustrated in a spectrum that ranges from the development of backyard habitat for songbirds and butterflies in suburban areas to the creation of an expansive network of breeding, migration, and wintering areas for migratory birds across the breadth of North America (e.g., whooping cranes and various waterfowl populations). That large-scale management is receiving greater focus is among the highlights of the studies appearing in this volume (e.g., the chapter on the American marten).

The growing store of evidence clearly indicates that our tools must be designed and used in ways that ensure the simultaneous accommodation of human activities and wildlife habitat. For example, the U.S. Fish and Wildlife Service released a plan in mid-1990 that addressed the long-term survival, genetic integrity, and recovery of the 30–50 endangered Florida panthers still remaining in the wild. The challenge will be to integrate panther habitat within a large area of southern Florida, where the accelerated growth of human populations has produced a multitude of urban, agribusiness, and recreational developments. Of the 2.2 million acres of occupied panther range, approximately 1 million are in private hands, where unmanaged activities have caused a significant loss of panther habitat during the past 25–50 years.

Maintaining the Florida panther as a viable component of a unique ecosystem remains a high priority for many Americans, who express a desire for expanding those programs that secure, restore, and improve the quantity and quality of panther habitat. In addition to acquiring more key tracts of land, attention will focus on working directly with private landowners to enlist their interest, participation, and commitment for preserving habitat. Cooperative agreements, leases, conservation easements, incentives, and compensation arrangements for landowners are among the tools awaiting design and application. While those initiatives deal with habitat quantity, other actions address ways that (1) risk factors might be minimized (e.g., road kills), (2) improve overall habitat quality, and (3) curtail human disturbances.

In other geographic areas, efforts focus on changing attitudes of people and moving toward large-scale management plans. The Greater Yellowstone Ecosystem in Montana and Wyoming covers approximately 31,000 square miles (19,900,000 acres). Diversity in topography and wildlife is nearly equaled by the multiple jurisdictions of government agencies and ownership of land. Sixty-nine percent of the ecosystem is publicly owned, with management carried out by five federal agencies (National Park Service, Forest Service, Bureau of Land Management, Fish and Wildlife Service, and Bureau of Reclamation). Private owners hold 24% of the area,

Indian reservations 4%, and states 3%. Management decisions made by any one of the landowners or agencies thus have implications for wildlife of the entire ecosystem. Wide-ranging species, such as elk and grizzly bears, roam 20 or more miles for food and cover and, at one time or another, occupy lands under various ownerships. These patterns of wildlife use, in combination with the drainage systems within the ecosystem, emphasize the essential nature of cooperative planning and management.

Although many of the remaining population of grizzly bears at times occupy federally owned wildlands, nearby ranches play an increasingly important role in the management of this endangered species. Some of the occupied bear habitat is privately owned land used for livestock ranching, and because they are located in relatively pristine and certainly attractive settings, ranches may be developed as sites for recreational, retirement, and other permanent homesteads—and with human encroachment, bears inevitably become a nuisance. Predictably, a call goes out for government action to eliminate the troublesome bears.

An obvious means of reducing these conflicts is to halt further development of homesites by supporting the livestock industry so that ranching remains a viable form of land use where it now exists in grizzly country. That calls for cooperative efforts among ranchers, recreationists, conservationists, and personnel from resource management agencies. Agencies should be prompt to remove troublesome bears. Ranchers can be paid for livestock lost to raiding bears, as some private conservation groups already do. Planned properly and carried out effectively, these actions would help perpetuate some of the area's unique values.

Although bear/livestock conflicts occur on ranches, studies have shown that most bears learn the ranch routine and habitually avoid the ranch personnel. Grizzlies generally stay to themselves and feed on wild plants and animals, often without anyone's knowledge that bears are around. These behaviors of bears provide promise for multiple benefits from large-scale planning and management in the Greater Yellowstone Ecosystem.

Similar needs for multiple-scale planning are evident for rangelands and associated riparian areas. Long-standing management efforts are being expanded to incorporate new visions and goals for more balanced uses of ranges. New, unprecedented alliances are now formed among federal and state agencies, livestock interests, fish and wildlife clientele, and the general public.

An important first step is to agree on goals, objectives, and procedures that are consistent in principle among agencies, regions, and states, thereby facilitating appropriate project-level decisions that are ecologically sound and economically reasonable. The U.S. Bureau of Land Management, for example, is developing a plan for 180,000 stream miles to improve the functioning and status of 23.7 million acres of riparian/wetland systems within the 270 million acres of public lands it administers. The goal is to improve habitat conditions to meet increasing demands for protecting watersheds, restoring water quality, and enhancing conditions for fish, wildlife, livestock, and outdoor recreational opportunities.

What too often remains poorly recognized is that ranching and farming, with appropriate management, can benefit certain species of wildlife. Some privately owned lands, for example, provide key wintering areas and crucial forage for elk and deer that spend other seasons on public lands. As is true for grizzly bears, elk, deer, and some other wildlife fare far better on rangelands rather than in developed areas subject to intense human activities and competition for space and water.

Obviously, ranchers and farmers must be involved with those planning efforts that concern overgrazing by livestock, restoring riparian/wetland systems, making available public recreational opportunities, and maintaining wildlife populations at desired levels. Properly managed ranges and riparian/wetland areas ultimately will produce desirable plant communities, replete with greater flexibility for realizing a greater variety of uses and benefits. With such efforts, hope remains that soil, water, and vegetation can be managed on a sustained basis to the advantage of both private and public sectors.

Large-scale cooperation and integrated management activities that extend over all land ownerships are required to perpetuate natural ecological processes and wild-living resources. Such joint efforts will require new levels of effectiveness. Most importantly, sound ecological comprehension must serve as the framework within which economic and political considerations and decisions occur. This framework is essential if we are to curtail the mounting costs for restoring damaged landscapes in the wake of unwitting, myopic, and/or ill-planned actions, including those that resulted from inappropriate political decisions.

A land manager, for example, focusing only on forest stands (each up to 100 acres) and compartments (each approximately 1,000 acres) may well miss the cumulative impacts on essential wildlife habitats over a larger area (tens of thousands to hundreds of thousands of acres). To avoid such disasters, multiple scales that include the habitat needs of wide-ranging wildlife species must be incorporated at the earliest stage of integrated planning for forest management. Planners and managers thus must be both nearsighted and farsighted when designing management prescriptions that assure sustainable forest ecosystems, including their wildlife populations.

Fragmented habitats may result from the absence of comprehensive, integrated resource management activities. The northern spotted owl, for example, in the Pacific Northwest requires stands of old-growth forests. A system of habitat conservation areas (HCA), each within 12 miles of another, and encompassing the home ranges of 20 pairs of owls, was recommended in mid-1990 by the Spotted Owl Scientific Committee. Under provisions of the plan, forests linking HCAs will be managed with an 80-year rotation, so, at times, 50% of the lands between HCAs have at least 40% forest canopy closure and trees with diameters of 11 inches or more. Full implementation of this management strategy enhances the probability that juvenile owls will disperse successfully, thereby maintaining the viability of spotted owl populations during the next century.

Other species of breeding birds also are threatened by the effects of forest fragmentation. Studies in Maryland, Michigan, and Oregon show that the occurrence of most species of forest-dependent birds is correlated with forest size.

Small islands of forest less than 33 acres are occupied by birds whose minimum territory sizes are met. These include edge species, permanent residents, and short-distance migrants that are granivorous or omnivorous in feeding habitats (e.g., jays, house wrens, cardinals, catbirds, chickadees, robins, starlings, and towees).

Islands of forest greater than 33 acres are required by nonpasserine carnivorous birds (e.g., hawks and vultures).

Large areas of contiguous forest of about 100 to more than 300 acres are required for an abundance of area-sensitive birds, primarily long-distance, insectivorous, neotropical migrants (e.g., flycatchers, vireos, and wood warblers).

The implications are clear: fragmentation of forest landscapes modifies the species composition of bird populations.

Without planning and with competitive bidding for development, landscapes with diversified plant communities invariably will be converted to urban/suburban areas dominated by buildings, concrete, blacktop, and exotic birds (e.g., starlings, house sparrows, and pigeons). With planning and minimal subdivision standards, habitats can be maintained and restored for such species as the cardinal, blue jay, house wren, catbird, robin, and tufted titmouse.

With overlay maps showing important characteristics of the landscape and wildlife areas, developers can design around the most sensitive and critical habitats and other ecological features. Such maps are available for all coastlines of the United States, some multicounty regional planning areas, cities, etc., but more maps need to be completed and used to accelerate integrated resource management.

Some progress has been made in developing a workable system of policies, guidelines, and procedures to integrate resource management more effectively. Further advances will require cooperative work among professionals of many disciplines (planners, landscape architects, hydrologists, soil specialists, biologists, resource managers, decision-makers, property owners, appointed and elected officials, and other citizens).

Each plan for units of the landscape, ranging from seasonal habitats among countries for migratory birds, to entire river or lake basins (e.g., Great Lakes), to suburban/urban areas, to homesite lots, should include prescriptions for wildlife habitats. By blending ecological and human needs, landscapes can be designed to perpetuate a broad spectrum of values and services from the resource base—frequently close to home—that will help provide humans with reasonable living standards. Polls repeatedly show the strong interests and desires of citizens to perpetuate landscape features that yield such qualities. The contributors to this important book thus provide timely insights on how to register significant accomplishments for the betterment of humans and the landscapes in which they dwell.

As emphasized by authors of this volume, human activities now dominate the landscape and will continue to do so in the future. Thus, it is essential to have well-designed plans and actions to ensure that human activities are compatible with maintaining the functions and health of ecosystems for all living resources. Integrated management approaches, such as those identified in this book, hold much promise for perpetuating ecological, economic, and social values and benefits. Simultaneously, these approaches will help to avoid costly restoration resulting from inappropriate plans and actions. The challenge is to see that well-designed, sensitive management approaches are applied broadly. People, as well as wildlife, will benefit immensely through the application of integrated resource management called for in this timely book.

—Laurence R. Jahn
President
Wildlife Management Institute

WILDLIFE
AND HABITATS
IN MANAGED
LANDSCAPES

Introduction

PROGRESS IN resource management is closely related to the continuous development of its separate disciplines. Virtually no advances in resource management could be made without the constant renovation of the various disciplines' knowledge bases and belief systems. Imagine wildlife biologists limiting their interest and knowledge in wildlife to the abstraction of the species or the individual. What progress in land-use planning would there be if the landscape architect's involvement stopped with the visual and spatial attributes of the master plan, or if the commercial logging company's interest for the forest stopped with the harvest?

During the 1989 North American and Natural Resources Conference, we explored various strategies for meeting natural resource needs. In the session "Wildlife and Habitats in Managed Landscapes," one strategy emerged, which suggested meeting resource needs by redefining habitats to include the concept of landscape. The concept, not landscapes as phenomena but as environments, has a significant meaning for the way we plan for people, habitats, and wildlife.

This strategy of resource management suggests applying the tools of planning, management, and design to the entire landscape itself, not merely to discrete entities in the environment. This concept of landscape involves resource managers in two kinds of spatial manipulations in which wildlife, habitats, and human aspirations are blended together. One kind of space requires all managers to view the landscape as a continuous entity. The second kind of space requires managers to organize the landscape's form and function for all parties concerned.

In times past, these two kinds of space were not mixed together successfully. There was an almost aristocratic concept that gave the managers all-powerful control over landscapes, and this view led to singlemindedness of purpose. Landscapes, therefore, were planned for the few in control.

Today, we see our landscapes as fields of conflict. Land-use decisions are

3

based upon legal and political decisions where compromises between authority and special interests are made. In such arenas, it becomes difficult to remember that the ultimate strategy of resource use must be to perpetuate the landscape, not to destroy it.

Resource-management improvements can be measured by how well we reconcile the differences between the pressures of human expansion and the limitations of our landscapes to produce natural resources. Three issues are central to this desired effect. First, resource managers must resolve the conflicts created in the spatial relationships among and between ecosystems. Second, resource managers must be cognizant of the landscape patterns we create over time as a result of our decisions. Third, resource managers must continually strive to develop innovative strategies that help to balance the energy and material exchanges between human land uses and the large landscape system.

Wildlife and Habitats in Managed Landscapes will focus on major papers that deal specifically with one or more of these resource-management issues. These papers have three characteristics in common:

Landscape Level of Resource Management.   The scale of a resource problem, be it a clearcut, an insect infestation, a collapse of a wildlife population, or an oil spill has always been much bigger in impact than our ability to respond to it. Foreman (1988), Fabos (1985), Lyle (1985), Odum (1971), Marsh (1983), and Harris (1984) each speak to the need of employing strategies, action packages of responses, and restoration work that is equal in magnitude and impact to the scale of disturbance.

The concept of an appropriate level of response was encouraged by compliance of federal resource managers with NEPA and Section 102(c), which requires federal agencies to identify impacts, unavoidable adverse effects, alternatives, long- and short-term trade-offs, and which makes an irreversible commitment of resources associated with federal land-use notions.

Now, predictably, the resolution of resource-use problems has slowly expanded to view the problem at a scale equal to its impact. In most cases, the spatial scale is at the landscape level, where patterns of land use, human agents, and impacted ecosystems are most appropriately assessed.

The Time Frame of Adequate Response.   In Rosenberg's (1988) report on the causes, aftermath, and control of greenhouse effects, he brings up the concept of adequate time frames for responding to this global crisis. He states that the more rapidly the climate changes, the more difficult it will be for natural ecosystems and human activities to anticipate and adapt to those changes. Viewed in the light of resource uses, all scenarios of habitat change imply a distribution of impacts, of costs and restoration efforts that demand a long-term, long-range recovery period. The best way we can respond to the magnitude of such impacts is to understand the time frames in which either preventive or restorative efforts would be applied to resolve them.

Technological Assistance.   Technological innovation is a primary reason we are able to think and act in large scales of space and across larger time frames. Were it not for the advent of remote sensing and, more specifically, the application of G.I.S. and landsat thematic mapping techniques, we could not resolve many of our present day resource-use problems.

The techniques of blood chemistry and urine analysis developed by DelGiudice, Mech, and Seal have given resource managers a solid basis for rethinking the value of winter habitat. Without the insights brought forth in this highly specialized work, we could do little more than hypothesize about the relationship of the nutritional status of deer and the food component of their winter habitat.

These technological advances demonstrate the necessity of merging the evidence of science with the intuition, or art, of resource management. These approaches advance us to the next order of evolution in our respective fields.

These three characteristics will no doubt become a more traditional part of the growing knowledge base and belief system of resource management.

Commonalities aside, what of the resource issues and, more importantly, what of our contribution to their resolution? Do these papers reflect progress and, if so, what impacts might they have?

THE SPATIAL RELATIONSHIPS BETWEEN ECOSYSTEMS

The projects described do contribute to the improvement in the use of our resources. Perhaps the most notable improvement may be seen in the types and kinds of spatial relationships we plan and design for our ecosystems.

Consider the marten project studied by Bissonette et al. (1989). They propose long-range landscape planning as a means to maintain both the survival of marten and old-growth forests in western Newfoundland. Certainly the timber industry and marten habitat could exist without the presence of the other. But in order for them to coexist, certain aspects of forest structure have to be preserved. The solution requires the preservation and integration of various seral stages and their spatial distribution across the broader landscape system.

Long-range planning for renewable resources takes on a different role in McComb's work in Kentucky (McComb et al. 1989). Coal revenues currently dominate the regional economy in the southern Appalachians and, since 1954, surface mining has impacted 10% of the commercial forestland in Kentucky. McComb suggests this dominant use of a nonrenewable resource threatens the future viability of the region in two very significant ways. First, surface mining irreversibly commits the entire landscape and its renewable resources to an activity dependent on a single, nonrenewable resource. Second, this activity threatens the remaining forest resources with increased management pressures.

McComb's recommendations include an analysis of the cumulative effects of the surface mining and forest management in this multiple-ownership region of Kentucky. At the very best, this activity may greatly improve the coordination of two very incompatible land-use activities. In so doing, a more rational approach to a very short-term economic decision may be established. Such a baseline could lead to a more acceptable reallocation of resources in the future.

THE IMPROVED LANDSCAPE PATTERN

A second benefit derived by these projects is the contribution they make to an improved landscape pattern in the greater landscape system. Land-use decisions, whatever their impacts, in time accumulate in the landscape pattern to become part of the overall fabric. These projects, if implemented, select for more favorable landscape patterns and, hopefully, better ecosystem performance. When land uses are not coordinated wisely or when they threaten productive landscape performances, an unfavorable environmental problem is perpetuated.

Such is the case in the rain forests of southeastern Alaska. Samson and his colleagues (1989) are studying ways to remedy the problems associated with resource utilization that impacts non-commodity values of wilderness, wildlife habitats, and biological diversity of forest ecosystems. Their work is directed at improving the means by which the forest is planned and harvested so as to improve yields over time as well as maintaining the landscape pattern.

Gates's work on corridor impacts is equally significant. He has developed baseline data that will yield insights into how to construct corridors that will improve wildlife habitats impacted by a major physical disturbance.

INNOVATIVE STRATEGIES

These projects give rise to strategies that might advance wildlife and their habitats in managed landscapes. Cain (1968) supported the notion that ecological principles have a direct correspondence to human ecosystems. He continually promoted the idea that conservation should be directed toward the interjection of ecological knowledge into human-action patterns. In this sense, ecological principles can be applied to the management of natural resources as well as to our manner of utilizing human resources. This unifying theme sets the stage for two subthemes.

Sustainable Landscape Mosaics

Harris (1984) and Foreman (1988) expand upon the notion that managers should build sustainable landscape mosaics. Dansereau (1957) further promotes this concept with his laws of community adjustment.

The concept of sustainable landscape mosaics is workable. It suggests the reformation of the yet-to-be-developed human-landscape mosaics in and around a skeletal corps of natural ecosystems where the former subordinates their activities. In so doing, structure and function of the area is given priority, and thereby a genuine social and institutional value. These landscapes become part of the perpetual landbank in which we can live and learn. DeGraaf is developing important information vital to the breeding bird populations in the logged forests of New England. He suggests that the only feasible way to achieve sustainable bird populations within these forests is to manipulate vegetation by selecting those key habitat elements essential to birds. This management strategy could, if carried out, benefit both the forest and the birds.

DelGiudice et al. (1989) offer a wealth of potential management insight with their nutritional analysis for mammals. This information makes it possible for resource managers to design a more appropriate palette of plant materials within the home range of a desired species of wildlife.

Bryant's extensive work on juniper woodlands fully recognizes the complex interrelationships between rangeland management and the vegetative communities in central Texas. He suggests a comprehensive management strategy that will sustain grazing, but at the same time accommodate the growing need to utilize these landscapes for residential, recreational, and wildlife values. This goal can be achieved by improving the way we plan, design, and manage the vegetative patterns of the landscape.

Such efforts are realistic and achievable since they engage each region and land ownership on their own terms. The authors have demonstrated how this can be accomplished on their specific projects. The concept also engages the right players, namely the land managers, the public, and the private enterprise system.

Regenerated Landscapes

One of the most certain impacts when landscapes are degraded is the associated reduction in quality of life and health for individuals. Predictions of atmospheric and water pollution, coupled with reports of increased crime rates and homelessness, lead us to believe the worst. Certainly these are grounds for concern. But there are also grounds for hope and constructive action.

Ulrich (1984) and McCarthy (1987) point out the many benefits that outdoor experiences have on individuals. We are discovering that the relationship between individual and landscape is a very real and vital part of everyone's cognizance.

Personal and life-long connection with landscapes is the key to promoting the regeneration of new landscapes. Using this concept, we can begin to retrofit parts of our human-dominated environments with less intensive uses where native flora and fauna still survive.

Regenerated landscapes are different from parks and open spaces in several ways. The diversity of the plant materials should reflect a naturally regenerating landscape. In fact, major portions should regenerate naturally, while some portions should be planted. The types of plants should be selected from plant palettes that are native and dominant in the area. Limited areas should be zoned for specific activities that are compatible with adjacent land uses.

Cable's concerns for the deteriorating condition of windbreaks and shelterbelts in the midwestern states demonstrate one example of such a concept. These landscapes, as vital as they are for wildlife habitat and recreational opportunities, are considered to be, at best, of marginal value to the dominant agricultural land uses. They need not be. Cable suggests that the regeneration of key habitats could augment the agricultural economy. The work required in many cases would be minimal, but could yield maximum benefits to the land owners and local inhabitants alike.

Szaro's work on riparian and stream ecosystems points out the need to reintegrate these landscapes into the land-use planning activities associated with these sites. These landscapes are, in many areas, the most vital and productive ecosystems available, yet they are not managed as such. Their continued exploitation without proper management and protection will guarantee their demise. Szaro has been studying ways to selectively protect and enhance their value so as to maintain riparian landscapes for future uses.

No single scenario of planting schemes or location should be dominant. The point in establishing these landscapes would be to build plant environments on lands where their presence would be guaranteed for at least a generation. Forest Service lands, B.L.M. lands, state forests, park lands, and even municipal open spaces might be primary candidates. The motive would be to build habitats on marginal lands where people and native wildlife could co-exist.

The ultimate purpose of regenerated landscapes and sustainable landscape mosaics is to elevate people's awareness of wildlife and wildlife habitats in managed landscapes to a new order of consciousness. The concept of landscape diversity is a key to maintaining and protecting desirable species. By managing for wildlife at the landscape level, we hope to derive benefits for wildlife, landscapes, and the human population.

—Jon E. Rodiek

Literature Cited

Bissonette, J. A., R. J. Fredrickson, and B. J. Tucker. 1989. American marten: a case for landscape-level management. Trans. N. Am. Wildl. and Nat. Resour. Conf. 54:89–101.

Cain, S. A. 1968. General ecology, human ecology and conservation. Dep. Conserv. Resour. Planning and Conserv., Univ. of Michigan, East Lansing.

Dansereau, P. 1957. Biogeography: an ecological perspective. Ronald Press, NY. 394pp.

DelGiudice, G. D., L. D. Mech, and U. S. Seal. Physiological status of white-tailed deer during winter and browse diversity. Trans. N. Am. Wildl. and Nat. Resour. Conf. 54:134–145.

Fabos, J. G. 1985. Land-use planning from global to local challenges. Chapman and Hall, NY. 223pp.

Forman, R. T. T. 1988. Ecologically sustainable landscapes: the role of spatial configuration in proceedings. Pages 11–26 *in* selected educational session 1988, IFLA World Congress. Landscape ecology and sustainable development session. Landscape/ Land-use planning. ASLA Washington, DC.

Harris, L. D. 1984. The fragmented forest. Univ. Chicago Press, Chicago. 211pp.

Lines, L., Jr., and L. D. Harris. 1989. Isolation of nature reserves in north Florida: measuring linkage exposure. Trans. N. Am. Wildl. and Nat. Resour. Conf. 54:113–120.

Lyle, J. T. 1985. Design for human ecosystems. Van Nostrand Reinhold, NY. 277pp.

Marsh, W. M. 1983. Landscape planning; environmental applications. Addison-Wesley Publ. Co., Menlo Park, CA. 356pp.

McCarthy, M. M. 1987. What would be the form of our cities and towns if living well really mattered? Landscape Australia 3/87:237–241.

McComb, W. C., K. McGarigal, J. D. Fraser, and W. H. Davis. 1989. Planning for basin-level cumulative effects in the Appalachian coal field. N. Am. Wildl. and Nat. Resour. Conf. 54:102–112.

Odum, E. 1971. Fundamentals of ecology, 3rd ed. W. B. Saunders Co., Philadelphia. 251pp.

Rosenberg, N. J. 1988. Greenhouse warming: causes, effects and control in renewable resources. J. of Wildl. Manage. 45:293–313.

Samson, F. B., P. Alaback, J. Christner, et al. 1989. Conservation of rain forests in southeast Alaska: report of a working group. Trans. N. Am. Wildl. and Nat. Resour. Conf. 54:121–133.

Ulrich, R. S. 1984. View through a window may influence recovery from surgery. Science 224:420–421.

© ED HOAG '90

1

Powerline Corridors, Edge Effects, and Wildlife in Forested Landscapes of the Central Appalachians

J. EDWARD GATES

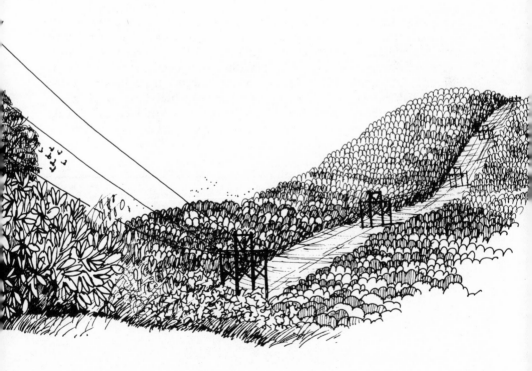

HUMAN DOMINATION of the landscape assumes a variety of forms. The process of urbanization produces perhaps the most noticeable forms. Associated with this process are the landscape conversions that result from connecting many urban centers scattered throughout a larger natural ecosystem. We are referring to the powerline, gas pipeline, and highway right-of-way corridors that crisscross the landscape.

When corridors are cut through forested landscape, such as those found in the central Appalachians, several changes occur simultaneously. First, there is a physically dramatic disruption to the continuous vegetative community. Second, there is a disruption to the structure and function of the wildlife habitats. Finally, there are impacts to the resident wildlife, which must negotiate, tolerate, or otherwise cope with these disturbances.

Urbanization fragments the natural ecosystem into smaller, less cohesive pieces of landscape. This fragmentation process raises many questions for the landscape manager concerned with wildlife. How do these large, spatial-scale landscape disturbances affect regional wildlife and habitat resources? Will wildlife be able to cope with this ever-growing process of reshaping the wild environment? Are there ways to optimally construct corridors in order to reduce the impacts they will have on the region's wildlife and habitat resources?

Research wildlife biologists have discovered that the predictability of these effects on wildlife habitats is not as optimistic as once believed. Dr. Gates focuses his attention on the effects of changes in the wildlife habitats found in and around corridors in the forested landscape of the central Appalachians. His studies lend insight to the various ways corridors function within the forest matrix as special habitats, as behavioral barriers, as filters to the movement of wildlife, and as a source of influences on the larger forest matrix.

The work represented by Gates and others has at least two important effects. First, it points out the need to conduct research that creates options for wildlife habitats formed within this new landscape mosaic. Second, it suggests that corridors may be a modern-day necessity in the central Appalachians. His work presents certain opportunities to design, plan, construct, and maintain corridors incorporating concerns for wildlife and habitats in this forest.

INTRODUCTION

Depending upon the surrounding background habitat or matrix, corridors in the landscape can be categorized as either of two major structural types. Convex corridors, such as hedgerows or forested stream corridors in agricultural areas, have greater height than the adjacent matrix. Concave corridors, on the other hand, are characterized by lower heights than the surrounding matrix. Powerlines, gas pipelines, and highway rights-of-way in forested areas are examples of concave corridors. Each type has unique properties and special characteristics that influence the local wildlife populations (Forman and Godron 1986).

Corridors function in a variety of ways within the matrix. For example, they may function as specialized habitats, as behavioral barriers or filters to the movement of wildlife, and as a source of effects on the matrix (Forman 1987). Furthermore, they may link together different patches in the matrix, providing travel lanes for animals between patches (Forman 1987, Harris 1988, Lines and Harris 1989). The way a corridor functions depends on several factors, including the animal species or species assemblage under consideration. Factors that influence how, when, and why animals use corridors include the composition and structure of the plant community within and adjacent to the corridor, the nature of the edge between the corridor and matrix, the size and shape of the corridor, and the microenvironment within the corridor, as well as whether the corridor is nested within other corridors. A highway corridor within a wooded corridor through an agricultural region functions much differently from a highway corridor through a forested region, even though forests border both highway corridors. The phenology and stability of the vegetation within corridors also influence animal use of corridors seasonally, as well as over successional time periods.

This study addresses two major questions related to powerline corridors and animal species or species assemblages in forested landscapes of the central Appalachians. First, how do the activity patterns of animals

15

differ among the major habitat types composing the corridor, edge, and forest? Second, what effects do specific corridor management practices have upon these activity patterns?

The Landscape Setting

The ridge and valley physiographic province of western Maryland was the location for the research. It is characterized by long, parallel ridges and intervening valleys oriented in a southwesterly to northeasterly direction. This region is bracketed by South Mountain in Washington County and Dans Mountain in western Allegany County. It contains strongly folded and faulted sedimentary rocks. The Great Valley in the eastern part of the region is formed on Cambrian and Ordovician limestone and dolomite. Farther west, a more rugged terrain has developed upon shale and sandstone bedrock that dates from the Silurian to Mississippian periods. There, some valleys are underlain by Silurian and Devonian limestones.

Within the western part of the region, two study sites, Green Ridge and Warrior Mountain, were selected along a 138 kV powerline corridor in Allegany County. This county has 83,603 ha, or 75.8% of the land area, in forestland. Of the productive forestland, 73.0% is in private ownership with most of the remainder in state forest, wildlife management areas, and parks. Two-thirds of the forestlands are classified as oak-hickory; a contiguous oak-hickory forest bordered both study sites. Overstory species included red (*Quercus rubra*), black (*Q. velutina*), and chestnut (*Q. prinus*) oaks; pignut (*Carya glabra*) and mockernut (*C. tomentosa*) hickories; and widely scattered pines (*Pinus virginiana* and *P. rigida*). The moderately dense (> 30% to 60% cover) understory was a mixture of oak, hickory, flowering dogwood (*Cornus florida*), red maple (*Acer rubrum*), and black gum (*Nyssa sylvatica*). Forest ground cover ranged from moderate (> 30% to 60%) to dense (> 60%), depending upon nearness to water, slope, or presence of canopy gaps. Species included seedlings of the understory and overstory trees, bracken fern (*Pteridium aquilinum*), common greenbrier (*Smilax rotundifolia*), and blueberry (*Vaccinium* sp.). This region receives 91.4 to 101.6 cm of rainfall annually.

GREEN RIDGE

This study site was located on the eastern slope of Green Ridge in Green Ridge State Forest. An herbaceous corridor was mowed with a rotary mower by personnel from the Maryland Forest, Park, and Wildlife Service on 5 August 1977, prior to and during the study on 18 July 1978, but not during the summer of 1979. Prior to mowing in 1978, the mean vegetation height was 27.6 cm (± 1.46 SE, N = 50); after mowing, vegetation height was reduced to 9.2 cm (± 0.83 SE, N = 25). Because of heavy rain, approximately 24.6 cm in July and August, the grasses in the corridor quickly recovered following mowing. The predominant grasses were orchard (*Dactylis glomerata*), redtop (*Agrostis alba*), and several panic (*Panicum* spp.) grasses. Forbs included spreading dogbane (*Apocynum androsaemifolium*), bracken fern, and lespedezas (*Lespedeza cuneata* and *L. bicolor*). Three lobes extending from the forest edge into the corridor were inaccessible to the mower and were dominated by blackberry (*Rubus allegheniensis*). The high-contrast linear edge was composed of seedlings and saplings of red, black, and chestnut oaks with a ground cover of orchard and redtop grasses, bracken fern, and small, scattered blackberry patches. The corridor width averaged 46 m (± 0.43 SE, N = 35). The site increased gradually in elevation from approximately 335 m to 427 m. It was crossed by one perennial stream and two intermittent tributaries. Soils included well-drained shaley silt loam and very stony loam.

WARRIOR MOUNTAIN

This study site was located on the eastern slope of Warrior Mountain in Warrior Mountain Wildlife Management Area. This shrubby corridor was managed for selected woody plant species by spraying with herbicides at 1- to 3-year intervals. Prior to the fieldwork, the most recent application occurred on 2 January 1978, when a mixture of 13.2 l of Tordon 155 and 365.3 l of no. 2 fuel oil was applied as a basal spray to certain target species. The most abundant shrub species was blackberry. Other species included staghorn sumac (*Rhus typhina*), mountain laurel (*Kalmia latifolia*), blueberry (*V. angustifolium* and *V. vacillans*), bear oak (*Q. ilicifolia*), and seedlings and saplings of sassafras (*Sassafras albidum*) and red maple. These species often were scattered throughout the corridor in small, irregular patches. Many small (< 8 cm dbh) snags were also present. Bracken fern, hayscented fern (*Dennstaedtia punctilobula*), whirled loosestrife (*Lysimachia quadrifolia*), and several species of panic grasses were the principal herbs. The two ferns often were found in large, homogeneous patches. Whirled loosestrife was prevalent throughout the corridor. The linear edge was of high contrast but less so than at Green Ridge. It was dominated by seedlings and saplings of red, black, and chestnut oaks, many of which later died from herbicide spraying. Large patches of blackberry and bracken fern extended from the edge into the corridor, providing some irregularity to an otherwise linear edge. The corridor averaged 52 m ($\pm$ 0.60 SE, N = 35) in width. The site increased rapidly in elevation from about 335 m to 610 m. It was crossed by one small intermittent stream. Soils were well-drained cherty silt loam and stony sandy loam.

THE APPROACH

Activity patterns of animals were determined by recording the number and location of crossings (i.e., a line of tracks left by an animal) made by different species or species assemblages over a 24-hour period on a sand-tracking surface measuring 0.61 m by 29.26 m (Bider 1968). Three such surfaces or transects were constructed at each site. A corridor transect was directly below, and parallel to, the center powerline; an edge transect, including both the corridor and forest edge, was perpendicular to, and bisected by, the forest boundary; and a forest transect was parallel to, and 60 m north of, the forest boundary. Most mammals were identified to species, but shrews were identified only to family. Other animal species were grouped according to lowest taxon (Table 1.1.). For example, birds were grouped into the Class Aves.

Transects were checked in early morning, and all tracks erased immediately afterward. Weather was noted after checking the transects and over the previous 24 hours. Each transect could be checked in 0.5 hour. Sampling periods covered 14 consecutive days in each of the months from May through August 1978 and 1979. Heavy rain in July 1978 resulted in data collection for only 10 days at Warrior Mountain and 13 days at Green Ridge. A transparent polyethylene plastic cover over a wooden A-frame helped to keep the sand-tracking surface dry.

Habitat-specific Activity Patterns of
Mammal Species and Species Assemblages

GREEN RIDGE

Because of overlap within the edge of species more characteristic of either the forest or corridor, richness was 1 to 3 species higher in the edge than within adjacent habitats (Fig. 1.1.). The mammalian assemblage had

TABLE 1.1.
Common and Scientific Names of Animal Species and Species Assemblages Mentioned in the Text

Common Name	Scientific Name
Mammals	*Class Mammalia*
Shrews, primarily short-tailed shrew	Family Soricidae, *Blarina brevicauda*
Red fox	*Vulpes vulpes*
Eastern chipmunk	*Tamias striatus*
Woodchuck	*Marmota monax*
Gray squirrel	*Sciurus carolinensis*
Red squirrel	*Tamiasciurus hudsonicus*
White-footed mouse	*Peromyscus leucopus*
Eastern wood rat	*Neotoma floridana*
Meadow vole	*Microtus pennsylvanicus*
Raccoon	*Procyon lotor*
Striped skunk	*Mephitis mephitis*
House cat	*Felis silvestris*
Dog	*Canis familiaris*
Opossum	*Didelphis virginiana*
Other animals	
Birds, primarily passerines	Class Aves, Order Passeriformes
Snakes	Order Squamata, Suborder Ophidia
Turtles, box and wood, primarily box turtle	Family Emydidae, *Terrapene carolina, Clemmys insculpta*
Salamanders/lizards	Order Urodela/Order Squamata, Suborder Lacertilia
Toads	Genus *Bufo*
Arthropods, primarily insects	Phylum Arthropoda, Class Insecta
Earthworms	Class Oligochaeta
Snails/slugs	Class Gastropoda

fewer crossings on the corridor transect compared with transects in the other two habitats (Table 1.2.). Mammal species or species assemblages were grouped according to their response to the three habitat types. Species or groups showing a filtering effect or avoidance of the corridor included the gray squirrel, red squirrel, white-footed mouse, and raccoon; avoidance of the forest was exhibited by the house cat and meadow vole, a potential prey species. A behavioral barrier at the corridor edge was shown by the eastern chipmunk in 1978 and 1979, and by the woodchuck in 1978. In 1979, the woodchuck showed similar activity in the edge and forest. This species inhabits open fields and forest edges. Shrews

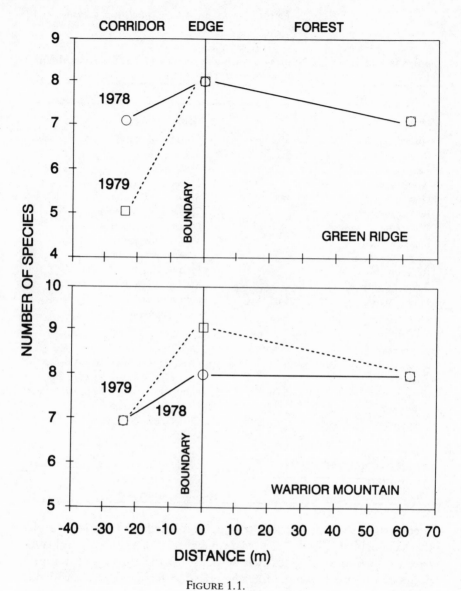

FIGURE 1.1.

Number of mammal species occurring in habitat types at each study site from late May through August 1978 and 1979. (Shrews, for which species were unknown, and feral dogs and cats are not included.)

were quite variable in their response to the habitat types from one year to the next. In 1978, there was no significant difference in number of crossings among habitat types, whereas in 1979 shrews showed a corridor avoidance response. This change could be related to differences in activity patterns from one year to the next or to changes in shrew species composition. However, most crossings were thought to be made by the common short-tailed shrew, especially within the corridor. Shrews may have been responding to movements by one or more prey species (e.g., snail/slug activity in 1979; Table 1.3.). The opossum was recorded on the study site in the second year. Although this forest species seemed to respond to the corridor as a barrier, the high activity within the edge could also have resulted from its use as a travel lane.

WARRIOR MOUNTAIN

Differences in species richness among the three habitat types in 1978 were slight with the forest and/or edge having one species more than the corridor. In 1979, the edge had one to two species more than the forest or corridor as a result of the additive effect of species from the adjacent habitats (Fig. 1.1.). The mammalian assemblage had lower numbers of crossings on the corridor transect compared with the other two habitats, and slightly more crossings on the edge (Table 1.2.). More activity occurred on the Warrior Mountain corridor transect than occurred at Green Ridge in both 1978 ($X^2 = 237.27$, df = 1, $P < 0.001$) and 1979 ($X^2 = 167.05$, df = 1, $P < 0.001$). Avoidance of the shrubby corridor was exhibited by the gray squirrel, red squirrel, eastern wood rat, raccoon, and dog. The gray squirrel responded to the corridor as a behavioral barrier, as did the raccoon in 1978. A wood rat den site was observed in a rocky ravine within the forest interior. Species that exhibited avoidance of the forest included the red fox and striped skunk. Both species prefer farmland, sparsely wooded areas, brushlands, and dense weed patches, usually near a stream or lake. These two mammals perhaps used the corridor and edge as travel lanes. Species responding to the corridor as a barrier to movement included shrews in 1979 and the eastern chipmunk in 1978. The movements of the eastern chipmunk significantly increased in the corridor from 1978 to 1979 ($X^2 = 395.42$, df = 1, $P < 0.001$) when greater vegetative cover likely provided more protection from hawks and other predators, as well as possibly increased foraging success. The fox, a potential predator of the chipmunk, also increased its use of the corridor in 1979. The shrubby corridor was less of an aversion to gray squirrels (1978:

TABLE 1.2.
Results of Chi-Square Tests of Crossings of Mammal Species and Species Assemblages by Habitat Type at Each Study Site from Late May Through August, 1978 and 1979

	1978				1979			
	No. Crossings	Forest (%)	Edge (%)	Corridor (%)	No. Crossings	Forest (%)	Edge (%)	Corridor (%)
Green Ridge								
Shrew	59	37.3	32.2	30.5	360	46.5	38.5	15.0**
Red fox	0	—	—	—	0	—	—	—
Eastern chipmunk	1341	37.9	61.1	1.0***	700	30.1	69.9	0.0**
Woodchuck	640	22.8	53.1	24.1***	1035	39.1	36.4	24.5***
Gray squirrel	222	66.5	32.5	1.0***	32	87.7	12.3	0.0**
Red squirrel	54	80.6	19.4	0.0**	9	100.0	0.0	0.0a
White-footed mouse	1163	70.4	29.6	1.0***	1688	45.4	32.3	22.3***
Eastern wood rat	0	—	—	—	0	—	—	—
Meadow vole	99	0.0	39.4	60.6**	71	0.0	4.3	95.7**
Raccoon	115	64.2	28.7	7.1***	104	53.5	24.1	22.4***
Striped skunk	8	13.1	63.5	24.4a	9	0.0	66.7	33.3a
House cat	134	0.0	11.3	88.7***	9	0.0	0.0	100.0a
Dog	1	100.0	0.0	0.0a	0	—	—	—
Opossum	0	—	—	—	18	33.3	66.7	0.0**
Total	**3836**	45.5	44.5	10.0***	**4035**	41.1	40.1	18.8***

Warrior Mountain

Shrew	20	20.0	40.0	40.0	73	33.3	53.3	13.4**
Red fox	0	—	—	—	154	21.2	38.6	40.2**
Eastern chipmunk	1253	38.1	54.1	7.8***	932	18.6	37.2	44.2***
Woodchuck	438	21.5	50.7	27.8***	1112	36.2	33.2	30.6**
Gray squirrel	217	35.9	41.0	23.1**	107	42.7	49.1	8.2***
Red squirrel	44	54.5	43.2	2.3**	12	8.3	91.7	0.0ᵃ
White-footed mouse	1588	31.4	31.2	37.4***	2111	30.4	38.4	31.2***
Eastern wood rat	197	93.2	6.8	0.0***	265	70.6	28.4	1.0***
Meadow vole	0	—	—	—	0	—	—	—
Raccoon	154	43.5	46.5	10.0***	451	39.5	32.3	28.2*
Striped skunk	29	3.1	40.3	56.6**	2	0.0	100.0	0.0ᵃ
House cat	0	—	—	—	0	—	—	—
Dog	2	50.0	0.0	50.0ᵃ	29	61.4	38.6	0.0**
Opossum	0	—	—	—	0	—	—	—
Total	**3942**	36.2	40.8	23.0***	**5248**	32.2	37.2	30.6***

ᵃ Inadequate sample size

* P <0.05

** P <0.01

*** P <0.005

TABLE 1.3.
Results of Chi-Square Tests of Crossings of Other Animal Species Assemblages by Habitat Type at Each Study Site from May Through August, 1978 and 1979

	1978				1979			
	No. Crossings	Forest (%)	Edge (%)	Corridor (%)	No. Crossings	Forest (%)	Edge (%)	Corridor (%)
Green Ridge								
Birds	848	23.7	43.3	33.0**	474	23.2	52.5	24.3**
Snakes	24	4.2	75.0	20.8**	15	6.7	86.6	6.7*
Turtles	25	48.0	32.0	20.0**	16	31.3	31.3	37.4
Salamanders/lizards	25	24.0	64.0	12.0**	2	0.0	100.0	0.0[a]
Toads	140	43.2	48.6	8.2**	260	8.1	61.5	30.4**
Arthropods	3843	30.3	30.6	39.1**	3196	33.3	40.5	26.2**
Earthworms	123	34.1	13.0	52.9**	152	28.3	13.1	58.6**
Snails/slugs	739	39.1	50.1	10.8**	982	57.5	35.1	7.4**
Warrior Mountain								
Birds	206	7.1	27.4	65.5**	250	1.1	30.8	68.1**
Snakes	10	20.0	30.0	50.0[a]	12	82.9	17.1	0.0[a]
Turtles	12	50.0	25.0	25.0[a]	7	0.0	0.0	100.0[a]
Salamanders/lizards	0	—	—	—	0	—	—	—
Toads	27	59.3	37.0	3.7*	223	34.5	32.3	42.2**
Arthropods	2650	58.5	25.0	16.5**	3547	34.6	39.1	26.3**
Earthworms	15	33.4	46.6	20.0*	27	67.2	22.2	10.6**
Snails/slugs	456	82.5	11.4	6.1**	854	72.6	22.1	5.3**

[a] Inadequate sample size
* P <0.01
** P <0.005

$X^2 = 51.52$, df $= 1$, $P < 0.001$; 1979: $X^2 = 2.88$, df $= 1$, $P < 0.05$) and white-footed mice (1978: $X^2 = 517.13$, df $= 1$, $P < 0.001$; 1979: $X^2 = 37.84$, df $= 1$, $P < 0.001$) than was the herbaceous corridor. Although white-footed mouse activity differed significantly among habitat types at Warrior Mountain, the percentage differences were slight and undoubtedly resulted from the large sample size.

Habitat-specific Activity Patterns of Other Animal Groups

GREEN RIDGE

The animal-tracking technique was effective only in sampling ground-foraging birds [e.g., the indigo bunting (*Passerina cyanea*), rufous-sided towhee (*Pipilo erythrophthalmus*), and field sparrow (*Spizella pusilla*)]. Birds had more crossings on the edge transect than on either the corridor or forest (Table 1.3.). Rather than functioning as barrier or filter, edge-adapted birds were merely using this specialized habitat for foraging. Snakes were also more active at the edge. Snakes included the timber rattlesnake (*Crotalus horridus horridus*), milksnake (*Lampropeltis triangulum triangulum*), black rat snake (*Elaphe obsoleta obsoleta*), black racer (*Coluber constrictor constrictor*), garter snake (*Thamnophis sirtalis sirtalis*), and hognose snake (*Heterodon platyrhinos*). The black rat snake tends to be associated with forest edges (Weatherhead and Charland 1985, Gibbons and Semlitsch 1987, Durner and Gates unpubl.). Snakes may have been responding to higher numbers of prey in the edge, based upon activity patterns of chipmunks, birds, and toads. Turtles, primarily box turtles (*Terrapene carolina*), appeared to avoid the corridor in 1978, but exhibited no difference among habitat types in 1979. Salamanders and lizards made more crossings on the edge transect. Toads were generally more active on the edge than in the forest or corridor. Arthropods, primarily ground-dwelling insects, centipedes, and millipedes, were quite variable in their use of habitat types from one year to the next. Activity in 1978 was slightly higher in the corridor, whereas in 1979 the edge had the higher number of crossings. Earthworm activity was highest in the corridor and then the forest. Snails and slugs in 1978 were more active in the edge; in 1979, this group was more active in the forest.

WARRIOR MOUNTAIN

Unlike at the Green Ridge site, birds at Warrior Mountain were more active in the corridor than in either the edge or forest, primarily because of the wider zone of shrubby habitat. Although foraging habitat was similarly available along the edge, birds tended to avoid the edge. Many bird species spent considerable amounts of time foraging in the corridor for seeds and invertebrates. There was no observable change in activity patterns in 1979, in response to further recovery of the corridor following defoliation. Toads in 1978 showed a corridor aversion, whereas in 1979, activity was higher in the corridor than in either edge or forest. This change could have resulted from more vegetative cover in the corridor, which provided greater protection from desiccation or predators. Arthropods, earthworms, and snails/slugs were least active in the corridor. Other than the hotter and drier microclimate within the corridor, some of these patterns could have been influenced by the activities of other groups of animals. For example, the foraging activities of birds and small mammals in the corridor could have resulted in lower numbers of arthropods, earthworms, and snails/slugs and, therefore, lower activity by these groups.

Powerline Corridors in Forested Landscapes: Their Functional Roles

SPECIALIZED HABITATS

How a powerline corridor functions depends on a number of factors that can be ascertained and evaluated. The animal species assemblage and the composition and structure of the corridor, edge, and adjacent habitat are the primary forces determining how a powerline corridor functions. Corridor width is often directly associated with the development of an interior environment within the corridor with interior species (Anderson et al. 1977, Johnson et al. 1979, Forman 1982, 1983). The contrast between Green Ridge and Warrior Mountain showed that this is not always the case. The Green Ridge and Warrior Mountain segments of the corridor were the same width. However, the first functioned as a borderline strip corridor with specialized interior habitat and species, whereas the second functioned more as a line corridor (Forman 1983). Furthermore, line corridors do not necessarily provide migration routes and habitats for edge species. At Warrior Mountain, mammal species occurring in the edge and corridor also usually were found within the forest interior. However, some species were characteristic of open fields and forest edges or of mixed habitats where trees were an important structural component.

At Green Ridge, the edge habitat or shrub border was only at the corridor-forest interface, whereas at Warrior Mountain it occurred throughout the corridor. Ground-foraging birds, several species of which were edge adapted, responded to this shrub zone by confining their foraging activities to the edge at Green Ridge. At Warrior Mountain, these birds occurred within the corridor interior, possibly avoiding high predation rates in the edge (Chasko and Gates 1982, Morgan and Gates 1982).

BARRIERS

A behavioral barrier occurs when movement of a particular species is normally concentrated within one of the two adjacent habitat types, and animal movements pile up at the edge between the two types (i.e., a "dam effect"; Jagomagi et al. 1988). Movement in the habitat on the other side of the barrier is minimal. An analogy could be that of a caged animal pacing back and forth along the perimeter of its cage. The tracks of such an animal would show that activity was concentrated along the perimeter. Few species or species assemblages demonstrated this trend on the two study sites. The composition and structure of the adjacent habitat types were critical factors in the functioning of edges as barriers. The dense, entangled mat of grass within the corridor at Green Ridge made it difficult to traverse for many small- or medium-sized mammals, resulting in avoidance by these species (Morgan and Gates 1983). However, comparisons with, and in consecutive years at, Warrior Mountain demonstrated that the barrier or aversion response at the edge can be collapsed by extending the width of the shrubby edge zone. Edge permeability increases as the contrast between adjacent habitat types decreases (Forman and Godron 1981). To increase it further would require succession within the corridor to proceed toward larger trees. The increased activity of many animal species within the shrubby corridor in year 2 indicated a deterioration of the barrier formed by the abrupt corridor-forest edge immediately following defoliation. If impermeable edges occur around patches, even small changes in edge permeability may have large impacts on animal movement across patch boundaries (Buechner 1987).

FILTERS

The majority of species responded to the corridor as an activity filter. This filter functioned to reduce activity with distance to the corridor (i.e., activity was highest in one of the adjacent habitat types, lower in the edge, and lowest in the habitat type not typical of the species). This was observed in both forest and grassland species. Corridor vegetation type could reduce the effectiveness of this filtering mechanism by essentially opening the pores of the filter.

TRAVEL LANES

The use of edge and corridor as travel lanes was observed in several medium-sized mammals that could easily have crossed the corridor, but instead moved along it more so than within the forest interior. Narrow multilayered edges support high densities of species and individuals because of their productivity and high contrast. Predator species may concentrate their activities in areas where their encounter rates with potential prey are high. Many predators are known to be more active along the edge because of the greater density and availability of prey (Gates and Gysel 1978, Chasko and Gates 1982, Forman 1987). However, an alternative hypothesis might be that the corridor edge represents the shortest distance between points A and B within the matrix or that it connects different resource patches (Wegner and Merriam 1979).

FLUXES

Forman (1982) noted that interactions among landscape elements involve flows of energy, nutrients, and species. Edges especially are thought to have major roles in species fluxes between patch and matrix. In fact, the main fluxes reported between corridor and patch and corridor and matrix have been species (Forman 1982). Nutrients and other materials are moved by foraging animals from edge to forest or corridor or across the boundary to, or from, adjacent habitats. If edges do indeed represent areas of greater animal concentration and activity, then high-contrast, narrow multilayered edges may function as catalysts for biological activity, thus enhancing the flow of energy between patches.

Powerline Corridor Management in Forested Landscapes: A Perspective

Although the central Appalachian region included in this study area is 78.5% forested, other land uses play important functional roles in the landscape. These include 4.6% cropland, 4.2% pasture, 4.3% other farmland, and 8.2% other land uses. Besides these patches within the forested matrix, a diverse network of natural and man-made corridors is present, ranging from wide highway and utility rights-of-way to narrow logging roads and hiking trails. These corridors fragment, as well as connect, different patches within the matrix. In a forested matrix, their ability to fragment habitat patches, isolate animal populations, and restrict gene flow are major concerns of ecologists and land managers. However, many of the detrimental effects of corridors, particularly man-made ones, within the forested landscape can be ameliorated by understanding the functioning of patches, corridors, edges, and matrix, and by proper planning and mitigation. With any newly designed or retrofitted powerline corridor, research is needed to ascertain that the alteration is functioning properly.

Powerline corridors could be feathered to avoid some edge effects. During construction, one or more clear-cut or selective-cut strips could be made along the corridor with the objective of creating a series of successional bands of vegetation parallel to the corridor opening. The first could be a shrub (< 2.54 cm dbh) strip, the next a strip of saplings ($\geq$ 2.54 cm, < 7.62 cm dbh), next a strip of pole-sized ($\geq$ 7.62 cm, < 30.48 cm in width) trees, and lastly the mature forest (> 30.48 cm dbh). These strips should be 10 m to 15 m wide, the width of many shrubby edge zones. Maintenance of these strips could be undertaken at the same time as that within the corridor. These strips should function to disperse both prey and predator. This would reduce encounter rates and the frequency of predation within edges and minimize the barrier effect found at abrupt edges. However, unless the corridor interior is altered, this procedure would do little to reduce isolation and increase gene flow. Furthermore, the wider edge would result in less forest interior for area-sensitive species.

Forman (1987) considered corridor width to be the primary control over the functioning of corridors. Width has a major influence on movement, i.e., the wider the corridor and the greater the contrast between corridor and the adjacent habitat, the more effective a barrier it becomes and the more likely the corridor interior will have a characteristic assemblage of animal species (Anderson et al. 1977, Johnson et al. 1979, Chasko and Gates 1982, and this study). Management procedures designed to reduce the effective width of the corridor should aid in collapsing the barrier effect. The response would vary with the animal species under consideration. As previously stated, a small change in an impermeable edge can produce a major change in movement across the edge. One beneficial procedure might be the creation of small lobes or peninsulas of shrubby vegetation extending from the forest edge into the corridor. These features tend to channel animals toward the tip of the peninsula where they may be more inclined to move in or out of patches (Forman 1987).

Another way animals can navigate past a barrier is through a breach or break. Powerline corridors traversing relatively flat terrain are difficult to design with such breaches. One way of incorporating breaches into the maintenance of the corridor is to alternate spraying of different sections of the corridor. Some sections, therefore, would have recently undergone defoliation, while others would be at a later seral stage. In hilly or mountainous terrain, the forest could be left intact where the powerline corridor crosses stream valleys or depressions, because transmission lines can be suspended above the vegetation. In fact, this is often the procedure in forested landscapes. Breaks in the corridor also would allow easy passage of many forest wildlife species from one side to the other, particularly those species with large home ranges. Breaks also could be important to dispersing or migrating wildlife species. However, this approach would not benefit mammals and other wildlife with small home ranges immediately adjacent to sections without such breaks, and may even create barriers to movement along the corridor. In order to create a breach for those species adjacent to the corridor, the best solution would be to establish a shrub community throughout the corridor. Such communities are initially more difficult to establish. However, once established, they tend to maintain themselves for long periods (Niering and Goodwin 1974, Gates: personal observation). Continuation of herbaceous corridors in the central Appalachians does not seem justified. They are used infrequently by most forest wildlife species, they function as a behavioral barrier or filter, and they require frequent mowing for maintenance and barrier reduction (Morgan and Gates 1983).

Breaches in the corridor also would be most critical at certain times of the year. Although this study did not monitor animal species assemblages throughout the annual cycle, it did provide data during the growing season. Increased crossings by mammals occurred on all transects during the spring/early summer and also later in the autumn. These peaks appear to be related to reproductive activities and dispersal of young. At such times, barriers often result in heavy mortality, as evidenced by the number of road kills observed along highways (Case 1978, Gates in press). Thus, efforts to reduce such barriers should be beneficial to the dispersal and survival of many wildlife species.

Acknowledgments

I thank Robert K. Rose, Ronald E. Barry, Jr., Steven W. Seagle, Frank C. Rohwer, and Donna M. Gates for reviewing drafts of this manuscript. I especially thank Anthony G. Ladino for collecting the field data, and Gregory G. Chasko and Stephen F. Oliveri for their assistance. Thomas P. Mathews and John E. Mash, Maryland Forest, Park and Wildlife Service, helped in the selection of study sites; Keith Martin, Potomac Edison Company, provided information on herbicide applications on the corridor. The Computer Science Center, University of Maryland, College Park, provided computer time. The study was made possible by Contract No. P46–78–04, Maryland Power Plant Siting Program. This is Contribution No. 2058–AEL, Center for Environmental and Estuarine Studies, The University of Maryland System.

Literature Cited

Anderson, S. H., K. Mann, and H. H. Shugart, Jr. 1977. The effect of transmission-line corridors on bird populations. Am. Midl. Nat. 97:216–221.

Bider, J. R. 1968. Animal activity in uncontrolled terrestrial communities as determined by a sand transect technique. Ecol. Monogr. 38:269–308.

Buechner, M. 1987. Conservation in insular parks: simulation models of factors affecting the movement of animals across park boundaries. Biol. Conserv. 41:57–76.

Case, R. M. 1978. Interstate highway road-killed animals: a data source for biologists. Wildl. Soc. Bull. 6:8–13.

Chasko, G. G., and J. E. Gates. 1982. Avian habitat suitability along a transmission-line corridor in an oak-hickory forest region. Wildl. Monogr. 82:1–41.

Forman, R. T. T. 1982. Interactions among landscape elements: a core of landscape ecology. Pages 35–48 *in* Perspectives in landscape ecology. Proc. Int. Congr. Neth. Soc. Landscape Ecol., Veldhoven, 1981. Pudoc, Wageningen, The Netherlands.

———— 1983. Corridors in a landscape: their ecological structure and function. Ekologia (CSSR) 2:375–387.

———— 1987. Emerging directions in landscape ecology and applications in natural resource management. Pages 59–88 *in* R. Herrmann and T. Bostedt-Craig, eds. Conf. proc., Science in the national parks. Vol. 1. U.S. Natl. Park Serv. and George Wright Soc., Fort Collins, CO.

Forman, R. T. T., and M. Godron. 1981. Patches and structural components for a landscape ecology. BioScience 31:733–740.

———— 1986. Landscape ecology. John Wiley and Sons, NY. 619 pp.

Gates, J. E., and L. W. Gysel. 1978. Avian nest dispersion and fledging success in field-forest ecotones. Ecology 59:871–883.

Gates, J. E. In press. Highways: the search for solutions. *In* S. S. Lieberman, ed. Conf. proc., Deer management in an urbanizing region: problems and alternatives to traditional management, April 13, 1988, Natl. Conf. Center, East Windsor, NJ. Humane Soc. U.S., Washington, DC.

Gibbons, J. W., and R. D. Semlitsch. 1987. Activity patterns. Pages 396–421 *in* R. A. Seigel, J. T. Collins, and S. S. Novak, eds. Snakes: ecology and evolutionary biology. Macmillan Publ. Co., NY.

Harris, L. D. 1988. Landscape linkages: the dispersal corridor approach to wildlife conservation. Trans. N. Am. Wildl. and Nat. Resour. Conf. 53:595–607.

Jagomagi, J., M. Kulvik, U. Mander, and V. Jacuchno. 1988. The structural-functional role of ecotones in the landscape. Ekologia (CSSR) 7:81–94.

Johnson, W. C., R. K. Schreiber, and R. L. Burgess. 1979. Diversity of small mammals in a powerline right-of-way and adjacent forest in east Tennessee. Am. Midl. Nat. 101:231–235.

Lines, Jr., L. G., and L. D. Harris. 1989. Isolation of nature reserves in north Florida: measuring linkage exposure. Trans. N. Am. Wildl. and Nat. Resour. Conf. 54:113–120.

Morgan, K. A., and J. E. Gates. 1982. Bird population patterns in forest edge and strip vegetation at Remington Farms, Maryland. J. Wildl. Manage. 46:933–944.

Morgan, K. A., and J. E. Gates. 1983. Use of forest edge and strip vegetation by eastern cottontails. J. Wildl. Manage. 47:259–264.

Niering, W. A., and R. H. Goodwin. 1974. Creation of relatively stable shrublands with herbicides: arresting "succession" on rights-of-way and pastureland. Ecology 55:784–795.

Weatherhead, P. J., and M. B. Charland. 1985. Habitat selection in an Ontario population of the snake, *Elaphe obsoleta*. J. Herpetol. 9:12–19.

Wegner, J. F., and G. Merriam. 1979. Movements by birds and small mammals between a wood and adjoining farmland habitats. J. Appl. Ecol. 16:349–358.

2

Windbreaks, Wildlife, and Hunters[1]

TED T. CABLE

[1] Kansas State University Agricultural Experiment Station contribution no. 90-177-B.

© ED HOAG '90

CONCERN FOR conservation of the landscape is not a recent human endeavor. The history of civilization bears evidence of the efforts made by users of the land to protect are: s whose values were seen as beneficial to people. One of the most pervasive examples of conservation over the centuries is the planting of windbreaks or shelterbelts to prevent soil erosion and to protect farmsteads from harsh winds.

In the United States, a national program of windbreak or shelterbelt plantings was started in 1934. The original purpose of this project was to provide protection for the agricultural soils susceptible to wind erosion throughout the prairie states. These plantings eventually provided much more than simple wind protection to the prairie ecosystems. They provided forested islands in a sea of agricultural land, where soil, moisture, plant communities, and wetlands could be sustained. As these landscapes matured, they also provided recreational opportunities, economic benefits from hunting and camping revenues, aesthetic values, and ultimately, visual relief from the prairie landscape itself.

Currently, as a result of intensive farming practices, many of these landscapes are suffering from age deterioration and a reduced size. Wildlife populations associated with these forested islands are being seriously affected. These forested islands, and the pressures to which they are being exposed, are the basis of Dr. Cable's investigations. He is concerned with the maintenance and protection of these island ecosystems, their contribution to the habitat opportunities for resident and migrant wildlife in this area, and the management of these habitats as a resource of regional significance.

Cable's work implies shelterbelts contribute significantly to the ecological processes and the life-support systems found in and around these landscapes. Consequently, the species diversity of the region and the availability of wildlife found there are linked directly to these man-made habitats.

He suggests that the future well-being of shelterbelts is related directly to two areas—the benefits derived from their protective functions and the economic values derived from the recreational activities they afford to the local inhabitants. Cable builds a case for developing data that will generate public support and political action favoring the protection of these island forests.

INTRODUCTION

Windbreaks or shelterbelts are tree plantings that impede wind flow. They consist of parallel rows of trees or shrubs oriented perpendicularly to the direction of the prevailing winds. At one time, the term windbreak described a tree planting designed to protect a farmstead or feedlot, whereas shelterbelt referred to a planting intended to prevent soil erosion in fields. However, the terms now are used interchangeably (Baer 1989).

This chapter presents a brief history of windbreak planting and reviews the literature pertaining to both wildlife populations in windbreaks and the design of windbreaks for wildlife. The chapter concludes with a socioeconomic study of the use of windbreaks by hunters.

The planting of windbreaks is not a new concept. The Scottish Parliament ordained the establishment of windbreaks in 1457. Russians began planting windbreaks in the 1700s. One of the earliest government-mandated programs for windbreaks occurred in Russia, where, in the early 1800s, Mennonites were required to spend 6 years planting trees in return for exemption from military service. When the Mennonites fled Russia in the 1850s, planting temporarily ceased. However, in the 1890s, a new initiative resulted in more than 130,000 ha of windbreaks being planted by the turn of the century. Throughout most of this century, the Soviets have maintained a strong commitment to windbreaks. A total of more than 2.5 million ha of windbreaks has been established to protect 19.8 million ha of agricultural land (Schroeder and Kort 1989).

The most extensive windbreak program in North America occurred on the Great Plains. Early settlers planted trees around their farmsteads for aesthetics and protection from the winds, but interest in windbreaks for crop protection was not widespread until the early 1900s. In 1934, President Franklin Roosevelt issued an executive order that led to the Prairie States Forestry Project in 1935. The idea was originally developed by

Roosevelt during a campaign trip in 1932, when his train was detained on a hot July day in a treeless area near Butte, Montana. This project became the largest windbreak program undertaken in the United States. Within eight years, 29,946 km (96,390 ha) of windbreaks were planted on 30,000 farms (Read 1958).

The agricultural benefits of windbreaks are well documented (e.g., Forman and Baudry 1984, Baer 1989). Windbreaks are used in virtually every region of the world where it is possible to grow trees (Brandle et al. 1988). At the International Symposium on Windbreak Technology held in Lincoln, Nebraska, in 1986, presentations detailed planting programs in 16 different countries (Hintz and Brandle 1986).

In the United States, various state and federal agencies offer free technical advice and financial subsidies to encourage landowners to plant windbreaks. Despite these incentives, the number of windbreaks on the Great Plains has been decreasing steadily over the past three decades. In a 13-county study area in Kansas, Sorenson and Marotz (1977) found that 20% of the windbreaks present in 1962 had disappeared by 1970. A study of five midwestern states showed that hedgerows—many functioning as windbreaks—were being removed at rates of 0.6% to 3.1% per year (Baltensperger 1987).

In addition to windbreaks being removed to facilitate modern farming practices and center-pivot irrigation, many are being lost to age-related deterioration. Sorenson and Marotz (1977) reviewed federal windbreak surveys conducted in 1947 and 1954, and concluded that significant deterioration had occurred during that interval. More recently, Schaefer et al. (1987) found that 61% of the windbreaks studied in South Dakota received a rating of *fair*, *poor*, or *no barrier.* A fair rating indicated that 30% to 40% of the canopy was missing, which resulted in a significant loss of effectiveness.

The demise of windbreaks on the Great Plains constitutes a loss of forest habitat that already is scarce in the region. Less than 3% of the Great Plains is forested (Griffith 1976), and the area is characterized by large expanses of unbroken monocultures of wheat or corn. Much of the native prairie and riparian habitat has been eliminated, leaving the few remaining windbreaks, hedgerows, and fencerows as the last vestiges of wildlife habitat. Although wooded habitats are scarce, Schwilling (1982) stated that woodland birds comprise the largest single component (46%) of the avifauna of western Kansas. Windbreaks on the Great Plains provide wooded habitat for wildlife species that otherwise would not be present

(Popowski 1976). If the trend of windbreak removal and deterioration continues, wildlife populations associated with these isolated tracts of wooded habitat will be negatively affected.

WILDLIFE IN WINDBREAKS: A LITERATURE REVIEW

In recent years, much attention has been given to the role of windbreaks as wildlife habitat. Windbreaks are thought to be responsible for the extension or maintenance of the ranges of fox squirrels (*Sciurus niger*), Mississippi kites (*Ictinia mississippiensis*) (Love and Knopf 1978), and other woodland bird species (Podoll 1979). Depending on the wildlife species and the season, windbreaks provide sites for reproduction, food, escape cover, and shelter from severe weather. The degree of importance of windbreaks varies by wildlife species. Of the 108 bird species listed by Johnson and Beck (1988) as using windbreaks, 29 species benefit substantially, 37 moderately, and 42 benefit only little or accidentally.

Johnson and Beck (1988) reviewed the literature and found references to 28 species of mammals using windbreaks. Of these, they considered the following 7 species as highly dependent on windbreaks in agricultural areas: eastern cottontail (*Sylvilagus floridanus*), desert cottontail (*S. auduboni*), gray squirrel (*Sciurus carolinensis*), fox squirrel, white-footed mouse (*Peromyscus leucopus*), southern red-backed vole (*Clethrionomys gapperi*), and European bank vole (*Clethrionomys glareolus*).

No large mammals are considered to be highly dependent on windbreaks. However, they do provide white-tailed deer (*Odocoileus virginianus*) with limited amounts of food, cover, (Podoll 1979) and fawning areas (Popowski 1976), and they may provide large mammalian predators (Order Carnivora) with travel corridors and hunting areas (Shalaway 1985).

Windbreak Functions

WINDBREAKS AS AREAS FOR REPRODUCTION

Most studies of wildlife in windbreaks have focused on breeding birds. Cassel and Wiehe (1980) identified the following five categories of use by birds during the breeding season:

1. Birds that carry on most of their activities within windbreaks, leaving only occasionally and for a short time. Black-billed cuckoos (*Coccyzus erythropthalmus*), least flycatchers (*Empidonax minimus*), house wrens (*Troglodytes aedon*), yellow warblers (*Dendroica petechia*), and northern orioles (*Icterus gulbula*) use windbreaks in this manner.

2. Birds that sing and nest in windbreaks and forage both in and out of windbreaks. This is the pattern of eastern kingbirds (*Tyrannus tyrannus*), western kingbirds (*T. verticalis*), American robins (*Turdus migratorius*), common grackles (*Quiscalus quiscula*), American goldfinches (*Carduelis tristis*), and song sparrows (*Melospiza melodia*).

3. Birds that nest and sing in windbreaks but forage widely. Mourning doves (*Zenaida macroura*) fall into this category.

4. Birds that use windbreaks for singing and forage in or out of windbreaks, but usually nest in adjacent fields. These include vesper sparrows (*Pooecetes gramineus*) and lark sparrows (*Chondestes grammacus*).

5. Birds that regularly use windbreaks for singing but rarely for foraging or nesting. Red-winged blackbirds (*Aeglaius phoeniceus*), dicksissels (*Spiza americana*), and meadowlarks (*Sturna* spp.) are examples of such species.

Moreover, in the Soviet Union, Kubantsev and Vasil'ev (1983) found that windbreaks also have a noticeable effect on the composition and numbers of breeding birds in adjacent fields, including species that have no relationship with windbreaks. This may be a result of the moderating effects as manifested in the vegetation type and condition, or it may be related to the increased populations of flying insects that occur on the leeward side of windbreaks (Lewis 1969) and serve as an atypically rich food supply for insectivores.

Studies of windbreaks during the breeding season have reported the following species-richness data for birds: 17 species nesting in 7 Minnesota windbreaks (Yahner 1982a); 29 and 36 species in 14 multirow and 14 single-row windbreaks, respectively, in South Dakota (Emmerich and Vohs 1982); 44 species in 69 windbreaks in South Dakota (Martin and Vohs 1978); 64 species in 81 North Dakota windbreaks (Cassel and Wiehe 1980); and 51 potential nesting species in 34 Kansas windbreaks during the reproductive season (Cable 1988).

Several studies have reported densities of breeding birds in windbreaks.

In South Dakota, Emmerich and Vohs (1982) found mean population densities of 3,306 individuals per 40 ha and 1,953 individuals per 40 ha for multirow and single-row windbreaks, respectively. Yahner (1982a) found 617 nests in 7 Minnesota windbreaks and reported an overall mean nest density of 88.5 nests per ha, with a range of 28.8 to 186.4.

The mourning dove is one of the most common birds associated with windbreaks and the one most studied. Mourning dove nest densities in windbreaks include 20.5 nests per ha (Yahner 1982a), 185 nesting attempts per ha, with 26% being successful (LaPointe 1958), and 33 nests (45 attempts) per ha (Randall 1955). Boldt and Hendrickson (1952) found an average of 7.5 nests per km of windbreak in North Dakota, whereas Wint (1978) reported more than 50 nests per km of windbreak in Oklahoma. Production estimates include: 84 doves produced per ha of windbreak (34 produced per ac) in Nebraska (LaPointe 1958); 69.7 doves fledged per ha in North Dakota (Randall 1955); and 9 young per km, also in North Dakota (Boldt and Hendrickson 1952). Yahner (1982a) and Weiser and Hlavini (1956) presented density data for other birds commonly nesting in windbreaks.

WINDBREAKS AS SHELTER

Windbreaks provide shelter from severe weather throughout the year. Ring-necked pheasants (*Phasianus colchicus*) use windbreaks for shade during warm, dry conditions (Hanson and Labisky 1964) and for shelter from heavy, spring rainstorms (Dalke 1943). For many species, the moderating effects of windbreaks may be most critical during winter, particularly in agricultural areas where little other cover exists.

The avian use of windbreaks during winter is sporadic. In North Dakota, Rotzien (1963) found 17 species in 8 windbreaks over 3 winters and noted that more birds used windbreaks during periods of severe weather. Cassel and Wiehe (1980) found only house sparrows (*Passer domesticus*) both breeding and wintering in the same windbreak. They concluded that, although several species may occasionally seek shelter or forage in windbreaks, no other species could be considered a winter resident in a single windbreak. Likewise, in South Dakota, Emmerich and Vohs (1982) found 9 species (mean density = 586 individuals per 40 ha) in 14 multirow windbreaks and only 2 species (mean density = 15 individuals per 40 ha) in 14 single-row windbreaks. In two winters, Yahner (1983a) found 13 and 16 species, respectively, in 7 Minnesota windbreaks.

Although most windbreaks contain few bird species at any one time during the winter, these places may serve as a refuge during severe weather, if they are properly designed and of sufficient width and height. However, bird use is difficult to predict based on short-term weather conditions (Yahner 1981) and is affected by the availability of other needs, particularly food, in or near the windbreak (May 1978, Stormer and Valentine 1981, Yahner 1981).

The moderating effects of windbreaks also play an important role in aiding thermoregulation of animals in adjacent areas during periods of cold wind (Johnson and Beck 1988). For example, Bennett and Bolen (1978) reported that windbreaks planted near small playa lakes on the Texas High Plains protected wintering ducks from winter winds, thereby enhancing the ducks' ability to conserve energy.

WINDBREAKS AS ESCAPE COVER

In light of recent studies of predator-prey relationships in edge habitat (e.g., Gates and Gysel 1978), the role of windbreaks in providing escape cover from predators is somewhat ambiguous. For a long time, escape cover has been considered a primary function of windbreaks benefiting wildlife populations. For example, the major determinant in the selection of roosting and loafing sites by pheasants seems to be protection from predators (May 1978). Specifically, windbreaks often provide areas of low herbaceous cover without a canopy (allowing birds to flush freely if a mammalian predator approaches) that are used by pheasants for night roosts and areas of dense, low canopy with little understory vegetation, which protects loafing birds from diurnal avian predators (May 1978). However, another hypothesis states that rather than providing refuge from predators the linear nature and small size of windbreaks may attract predators, either to use as a travel corridor or to search for prey that may be more concentrated than in larger habitats (Yahner 1983a).

Gates and Gysel (1978) observed unusually high levels of nest predation and brown-headed cowbird (*Molothrus ater*) parasitism along man-made edges, a phenomenon they called the *ecological trap hypothesis*. Because of their narrowness, most windbreaks consist almost entirely of edge habitat, and with few other trees in the area, windbreaks may attract avian predators. Ironically, for this reason, some windbreaks planted to increase game-bird populations may produce negative effects (Petersen 1979, Snyder 1985, Potts 1986, Hudson and Rands 1988, Carroll 1989). However, high rates of productivity have been reported in windbreaks for species that do not nest on the ground (e.g., LaPointe 1958), and Shalaway

(1985) reported higher nest success in fencerows than in native shrub or woodland. Basore et al. (1986) found no difference in nest predation in strip cover compared to agricultural fields. Shalaway (1985) found that in fencerows nest success was higher for passerine species (both ground and arboreal nesters) than for larger ground-nesting birds, such as mallards (*Anas platyrhynchos*) and ring-necked pheasants. Johnson and Beck (1988) reviewed this sometimes conflicting literature and concluded that a better understanding is needed of preditor-prey relationships in windbreaks.

WINDBREAKS AS FEEDING AREAS

Food availability in or near windbreaks is critical for most wildlife (Podoll 1979, Stormer and Valentine 1981, Yahner 1983*a*). Yahner (1983*a*) characterized bird species nesting in Minnesota windbreaks as predominately omnivores and granivores. Martin and Vohs (1978) reported that 34 of the 44 species breeding in windbreaks (77%) and 64 of the 68 migratory species (94%) feed entirely or partially on insects. Furthermore, the number of species and individuals that were completely or partially insectivorous increased as the size of the windbreak increased. In the 23 smallest windbreaks in their sample, 65% of the species were insectivores, whereas in the 23 largest windbreaks, 75% of the species were insectivores.

Lewis (1969) found that flying insects accumulate in the sheltered areas near windbreaks, especially on the leeward side. Barn swallows (*Hirundo rustica*), purple martins (*Progne subis*), chimney swifts (*Chaetura pelagica*), and other insectivores often are seen foraging on the leeward side of windbreaks, presumably taking advantage of the concentration of flying insects (Cable 1988).

OTHER WINDBREAKS FUNCTIONS

Although windbreaks are typically wooded islands in a sea of cropland, they also can serve as travel corridors to food resources in adjacent fields (Stormer and Valentine 1981). Moreover, if many windbreaks exist in an area, they can serve as stepping stones for migrating birds or those dispersing between riparian habitats and other wooded tracts (Yahner 1983*a*).

Migrant birds use windbreaks for feeding and resting. Martin and Vohs (1978) reported that during migration 68 species used windbreaks in South Dakota. Densities during spring migration were 2,389 individuals per 40 ha in multirow windbreaks and 2,906 individuals per 40 ha in single-row windbreaks (Emmerich and Vohs 1982). In 7 Minnesota wind-breaks, Yahner (1983*a*) found 39 species during fall migration and 64 species during spring migration.

WINDBREAK DESIGN AND MANAGEMENT IMPLICATIONS

Schroeder (1986) developed a model, as part of the U.S. Fish and Wildlife Service's Habitat Suitability Index (HSI) Model Series, that predicts avian species richness in windbreaks. This model includes six characteristics that should be considered when the objective is to create wildlife diversity: windbreak area, number of rows, plant diversity, windbreak height, canopy closure, and configuration.

Area.   The number of wildlife species in windbreaks is strongly corre-lated with windbreak area. The number of breeding birds in North Dakota (Cassel and Wiehe 1980), South Dakota (Martin and Vohs 1978), and Kansas (Cable 1988); spring migrants in South Dakota (Martin and Vohs 1978); and small mammals in Minnesota (Yahner 1983*b*) were strongly correlated with windbreak area. Martin (1981) concluded that as area increases the space requirements for additional species become avail-able. However, when minimum space requirements are met, the com-bined effects of habitat, chance, and competitive interactions influence the presence or absence of individual species.

Number of Rows.   Wildlife diversity increases as the number of rows increases (Cassel and Wiehe 1980). The number of rows may be especially important in winter. Podoll (1979) reported that snowdrifts often pene-trate 30.5 m into windbreaks; therefore, narrow windbreaks may lose their wildlife values. Some windbreaks may serve as winter death traps for pheasants if they are too narrow (Popowski 1976, Johnson and Beck 1988). The number of rows necessary for winter cover varies with lati-tude. In the southern Great Plains 4 or 5 rows seem sufficient (Capel 1988), whereas Yahner (1983*a*) suggested that windbreaks with 8 rows are desirable in Minnesota, and 20 rows may be necessary in northern regions (Capel 1988).

Plant Diversity.   In constructing the HSI model, Schroeder (1986) assumed that 6 or more species of woody plants in each windbreak represents an optimum for wildlife. He suggested that increases in woody-plant diversity increase the structural complexity of windbreaks.

Yahner (1982*b*) found that significantly more birds than expected were observed in the plant genera *Picea, Ulmus, Lonicera, Populus,* and *Acer.* Martin and Vohs (1978) evaluated the benefits of the trees and shrubs in South Dakota windbreaks. They found that open-foliage woody plants were most attractive to birds and that Siberian elm (*Ulmus pumila*) was the tree most widely used by birds. Hackberry (*Celtis occidentalis*) and green ash (*Fraxinus pennsylvanica*) also were positively associated with avian species richness.

Alternating tall deciduous trees with leguminous shrubs encourages diversity in single-row windbreaks. The choice of specific species varies with climate and site conditions; however, fast-growing and long-lived species are desirable for obvious economic reasons. From the perspective of wildlife management, species should be selected for their ability to provide food and shelter throughout the year. For example, some species of conifer should be planted on the windward side where they provide winter cover and impede snow penetration. Stormer and Valentine (1981) and Capel (1988) make specific recommendations for species that are particularly beneficial for wildlife in windbreak habitats.

Windbreak Height. Emmerich (1978) found that windbreak height was positively correlated with avian diversity, and in single-row windbreaks, height was the most important variable in determining diversity during both spring migration and the nesting season. Schroeder (1986) concluded that an average height of 16.5 m or greater was optimum for avian diversity, which would decrease linearly as the average height decreased.

Canopy Closure. Percent canopy closure was positively correlated to species richness in Minnesota (Yahner 1983*a*) and South Dakota (Martin 1978). However, certain species are negatively correlated with canopy closure (Martin 1978). Based on a review of literature, Schroeder (1986) concluded that canopy closure of 50% to 70% is optimum and that suitability for wildlife decreases linearly as percent canopy closure decreases to zero or increases from 70% to 100%.

Configuration. According to Schroeder (1986), the most-desirable configuration for wildlife is at least one row of shrubs on each side of internal rows of trees. The second most-desirable configuration is a shrub row on only one side. The third most-desirable configuration is trees only or trees on outside rows. The least-desirable configuration is a windbreak consisting of only shrubs. Outside shrub rows are valuable because they impede snow penetration, and on the leeward side, they often provide food near the ground in an area that is protected from the wind and where snow may not accumulate. Because the leeward side is generally on the south or east, it receives warmth from direct sunlight. A combination of outside rows of shrubs and internal rows of trees increases vegetative complexity, which increases wildlife richness.

OTHER FACTORS

In addition to the six characteristics identified in the HSI model, landowners can manage for wildlife by giving consideration to the following factors:

Presence of Snags.   Several studies have shown a positive correlation between windbreak age and bird diversity (e.g., Martin 1978, Cassel and Wiehe 1980, Yahner 1983a). Wildlife is probably not responding directly to age, but rather the age variable is reflecting an increase in snags, as well as increased tree height, area, and canopy closure. Snags provide foraging and nest sites for many species of birds and some mammals (Martin and Vohs 1978, Stormer and Valentine 1981, Yahner 1983a). The lack of snags may limit distributions and densities of cavity-nesting species, such as black-capped chickadee (*Parus atricapillus*) and downy woodpeckers (*Picoides pubescens*) (Yahner 1983a). The number of snags per hectare was strongly correlated with species richness (Cable 1988). However, as snags increase, noncavity nesting species may be adversely affected by changes in vegetative structure, such as a decrease in canopy closure.

Adjacent Feeding Areas.   Food availability can be critical for species that forage outside the windbreak (Lyon 1959, Cassel and Wiehe 1980, Stormer and Valentine 1981, Yahner 1982a, Love et al. 1985). Croplands, particularly grain fields, are more beneficial than grazed pasture (Yahner 1983a). Autumn plowing should be avoided, whereas alternative cropping systems, such as conservation tillage (especially no tillage) and organic farming near windbreaks, benefit wildlife (Yahner 1983a, Basore et al. 1986). Food plots placed near windbreaks are most beneficial if placed on the east or south sides to ensure access during the winter months.

Grazing.   Although periodic light grazing of a windbreak may increase vegetative complexity, thereby benefiting wildlife (Martin and Vohs

1978), grazing levels are generally too high to maintain understory vegetation (Stormer and Valentine 1981, Yahner 1983a). The understory is especially important for birds that nest or feed near the ground, and it is critical for providing these species with shelter from winter weather (May 1978).

Each of these factors has important benefits for wildlife, but some may not be controllable or economically feasible for landowners. However, if the objective is to promote wildlife richness and density, landowners or managers should consider as many of these factors as possible.

SOCIOECONOMIC BENEFITS OF WILDLIFE IN WINDBREAKS

Because many of the species associated with windbreaks are game species, windbreaks provide an array of hunting opportunities that are often scarce in intensively farmed regions. Research that evaluates the worth of the wildlife associated with windbreaks was recommended as a priority by the International Symposium on Windbreak Technology (Brandle et al. 1988). To address this concern, a study in Kansas assessed the amount of hunting done in windbreaks, the species of game involved, and the economic value of hunting in windbreaks.

Methods

A 12-page questionnaire based on the Total Design Method (Dillman 1978) was sent to 1,501 randomly selected, licensed, resident hunters in Kansas. One hundred and fifteen questionnaires were returned as undeliverable. Of the remaining 1,386 questionnaires, 842 were completed and returned, resulting in a 60.8% response rate. A telephone survey of 115 nonrespondents showed no evidence of nonresponse bias. Statistical analyses were carried out on SPSSPC+ V3.0 (Norusis 1988).

The survey instrument examined general hunting behavior, use of windbreaks, hunting-related expenditures, economic value of hunting in windbreaks, windbreak design, personal involvement in hunting and specialization, and demographic characteristics of the hunters. The questionnaire also collected specific information about the 1987–1988 hunting season.

The Contingent Valuation Method (CVM) assessed the economic value of hunting in windbreaks. This method creates hypothetical markets to simulate real-market conditions, thereby obtaining a price for a product. Consumers provide estimates of what they would be willing to pay for the product, and respondents bid on various quantities of the product in question. Federal agencies use CVM for determining the recreational benefits of federal water projects (U.S. Water Resources Council 1983). Sorg and Loomis (1984) and Walsh (1986) have reviewed CVM applications, and Cummings et al. (1986) have addressed concerns about the validity and accuracy of CVM and current CVM techniques.

Properly constructed CVM questions yield accurate measures of willingness to pay for hunting, and replication of several CVM studies produce consistent results for hunting values (e.g., Hammack and Brown 1974, Cocheba and Langford 1978, Bishop and Heberlein 1984, cited in Steinhoff et al. 1987).

Results and Discussion

HUNTER USE OF WINDBREAKS

Kansas hunters spend an average of 40.7% of their hunting time in or adjacent to windbreaks, and 81.2% of the respondents hunt in windbreaks at least some of the time. Some hunters (6%) reported that they spend 100% of their hunting time in windbreaks. Those who hunted in windbreaks made an average of 9.3 trips to their favorite windbreak site during the hunting season. Based on these data, windbreaks in Kansas support 1,370,000 hunter-days annually. In addition, latent demand may exist for windbreak hunting; 80.4% of the respondents said they would hunt in windbreaks more often if more were available.

Lyon (1961) found that hunters killed more pheasants in windbreaks, and with less effort, than in other habitats. High rates of success in windbreaks produce quality hunting experiences. Just as the predator-trap hypothesis suggests that predators are attracted to windbreaks, hunters likewise appear to exploit these areas.

Hunters who spent more than 75% of their time pursuing one type of game were categorized into groups based on their quarry (e.g., pheasant hunters). The mean percentage of hunting time spent in windbreaks was 55.2% for quail hunters, 40.0% for deer hunters, and 22.8% for pheasant hunters. T-tests indicated that each of these means are significantly different ($P < 0.05$ level).

The use of windbreaks by hunters from 60 western Kansas counties was compared with hunters from the remaining 45 eastern counties. The average percentage of time hunting in windbreaks by western hunters was 34.3%, whereas the average for eastern Kansas hunters was 45.2%. The difference is statistically significant ($P < 0.001$).

When hunting in windbreaks, Kansas hunters spent 29.4% of their time hunting quail, 28.7% for pheasants, and 13.7% of their time hunting deer. The following game accounted for the remaining time: doves (9.1%), rabbits (9.0%), squirrels (3.2%), coyotes (2.7%), turkeys (1.3%), prairie chickens (1.2%), waterfowl (1.1%), and other game (0.6%).

EXPENDITURES ASSOCIATED WITH HUNTING IN WINDBREAKS

The 1,370,000 hunter-days spent annually in windbreaks by Kansas hunters not only provide enjoyment and an improved quality of life for many hunters, but they also contribute to the state's economy. Expenditures often have been used as a measure of the economic importance of hunting (Steinhoff et al. 1987); however, expenditures are not a good measure of economic value. For example, a hunter might hunt free of charge on his or her own land, but the experience might have value. Although expenditure data do not measure the benefits the consumer gains by having the opportunity available, they are important measures of income transfers in the local economy (Sorg and Loomis 1984).

In Kansas, the average distance from the hunter's home to the most frequently hunted windbreak site was 71.9 km, and the average total cost for gas, food, and lodging was $19.97 per trip. For the 17.5% of hunters who took overnight trips to the windbreak site, average distance was 249.2 km, and average cost was $74.38 per trip. For the remaining 82.5% of the hunters who took day trips, the average distance was 34.3 km, and average cost was $8.42 per trip. The average day-trip hunter made 9.9 trips to windbreaks during the hunting season, whereas the average overnight hunter made 6.6 trips. These differences are statistically significant ($P < 0.01$).

The U.S. Fish and Wildlife Service estimates that resident hunters in Kansas spend $74,939,000 annually on hunting. Of this, $28,120,000 (37.5%) represents in-state, trip-related expenditures. Nonresident hunters spend an additional $13,502,000 annually in Kansas on trip-related goods and services (U.S. Dept. Interior 1988). Based on this estimate, residents hunting in windbreaks in Kansas spend $30.5 million annually. If nonresident hunters responded similarly, they spend $5.5 million annually in Kansas.

If all windbreaks were removed, these expenditures would not be lost. Many hunters undoubtedly would use substitute sites. However, loss of windbreaks would decrease habitat that, in turn, could decrease both the quantity and quality of hunting experiences. The large number of hunters reporting that they would hunt in windbreaks more often if more were available may indicate that windbreaks for hunting are already scarce. The lower percentage of time spent hunting in windbreaks by western Kansas hunters may reflect that opportunities are presently not available there.

CONTINGENT VALUATION ANALYSIS

Many hunters (84%) felt the last trip they took to their favorite windbreak site was worth more than they actually spent. When asked how much more they would be willing to pay in increased travel costs before they would not hunt again at their favorite site, bids ranged from $0 to $1,000, with a mean response of $25.46. Day-trip hunters' mean bid was $18.12, whereas overnight hunters' mean bid was $56.51.

Thirty-one respondents (6.5%) bid $0 for the additional travel cost they would be willing to incur. Because no question was included to determine the reasons for such bids, a t-test was run on the means with and without the $0 bids. No significant difference ($P < 0.01$) was found.

Bids of more than $100 also were screened. These bids were either accepted or rejected based on the hunter's trip cost, distance to the site, and length of stay. Again, no significant difference ($P < 0.01$) was found between the mean bids with and without the high bids. Therefore, the all-inclusive bid ($25.46) was considered a reliable estimate of the willingness to pay. On a hunter–day basis, the $18.12 per hunter–day for day-trip hunters and the $26.91 per hunter–day ($56.51 per 2.1 days per trip) for overnight hunters are well within the range of values reported in other studies (e.g., Hammack and Brown 1974, Bishop and Heberlein 1979, Sorg and Loomis 1984, Donnelly and Nelson 1986).

When hunters were asked how much more they would be willing to pay if they knew their game harvest would double, the values again ranged from $0 to $1,000, with a mean response of $28.52. Mean bids of $21.63 and $59.81 were recorded for day-trip and overnight-trip hunters, respectively. The same procedures concerning zero and high bids revealed no significant differences ($P < 0.01$). The bids associated with doubling of game harvest increased the average value by $3.51 per trip for day hunters and $3.30 per trip for overnight hunters. Statewide, this increase in harvest would add $3.3 million per year to the net economic value of windbreaks.

The CVM bids per hunting trip for current hunting conditions can be expanded to determine the net value of windbreaks for hunting in the entire state. Multiplying by the average number of trips per year to windbreaks, and expanding to the entire hunter population, yields a total net economic value of $14,978,771 for day-trip hunters and $6,559,285 for overnight hunters. Therefore, the total net economic value is $21,538,056 per year.

This figure may be conservative for two reasons. First, the values are based on the windbreak site most often visited by the respondent. The calculated estimates would be low because other, less frequented windbreaks also have some value to hunters. Second, the CVM value presented here represents the value only to licensed resident hunters. Out-of-state hunters account for 10% of the total hunter-days in Kansas, and there are 31,000 unlicensed hunters under the age of 16 in Kansas (U.S. Dept. Interior 1988). The values that these two groups place on hunting in windbreaks are not included in the CVM analysis.

Implications

The results of this study illustrate the importance of windbreaks as a recreational resource in Kansas. Virtually every type of hunting that occurs in Kansas takes place in or adjacent to windbreaks at some time. Windbreaks are particularly important in providing high-quality opportunities for quail, pheasant, and deer hunters.

Also, hunting opportunities provided by windbreaks have economic value, and a change in the quality or availability of these opportunities could negatively affect local economies in rural communities now dependent on hunters' dollars.

Windbreaks continue to decline on the Great Plains despite widespread recognition of their agricultural benefits. If this trend continues, wildlife populations may decrease, and both the quality and quantity of hunting opportunities will be affected negatively. As the number of windbreaks decreases, there will be less game and potentially more hunters concentrated in the remaining habitat. This will result in a diminished quality of the recreational experience; hunters may take fewer trips or even drop out of hunting. In turn, this has important economic ramifications, and it would affect the quality of life for many Kansas hunters.

Literature Cited

Baer, N. W. 1989. Shelterbelts and windbreaks in the Great Plains, J. For. 87:32–36.

Baltensperger, B. H. 1987. Hedgerow distribution and removal in nonforested regions of the Midwest. J. Soil and Water Cons. 42:60–64.

Basore, N. S., L. B. Best, and J. B. Wooley. 1986. Bird nesting in Iowa no-tillage and tilled cropland. J. Wildl. Manage. 50:19–28.

Bennett, J. W., and E. G. Bolen. 1978. Stress response in wintering green-winged teal. J. Wildl. Manage. 42:881–886.

Bishop, R. C., and T. A. Heberlein. 1979. Measuring values of extramarket goods: Are indirect measures biased? Am. J. Agric. Econ. 61:926–930.

————. 1984. Contingent valuation methods and ecosystem damages from acid rain. Dept. Agric. Econ. Staff Paper 217., Univ. Wisconsin, Madison.

Boldt, W., and G. O. Hendrickson. 1952. Mourning dove production in North Dakota shelterbelts, 1950. J. Wildl. Manage. 16:187–191.

Brandle, J. R., D. L. Hintz, and J. W. Sturrock, Editors. 1988. Windbreak technology. Elsevier Science Publ., Amsterdam, The Netherlands. 598pp.

Cable, T. T. 1988. Windbreaks as breeding habitat for nongame birds in Kansas. Kansas State Univ. Dept. Forestry Res. Rep. No. 8, Manhattan. 27pp.

Capel, S. W. 1988. Design of windbreaks for wildlife in the Great Plains of North America. Agric. Ecosystems & Envir. 22/23:337–347.

Carroll, J. P. 1989. Ecology of gray partridge in North Dakota. Ph.D. Thesis, Univ. North Dakota, Grand Forks. 145pp.

Cassel, J. F., and J. M. Wiehe. 1980. Uses of shelterbelts by birds. Pages 78–87 in R. M. DeGraaf, tech. coord. Management of Western forests and grasslands for nongame birds. U.S.D.A. For. Serv. Gen. Tech. Rep. INT–86.

Cocheba, D. J., and W. A. Langford. 1978. Wildlife valuation: the collective good aspect of hunting. Land Economics 54:490–504.

Cummings, R. G., D. S. Brookshire, and W. D. Schulze, Editors. 1986. Valuing environmental goods: an assessment of the contingent valuation method. Rowman & Allanheld, Totowa, NJ. 270pp.

Dalke, P. D. 1943. Effect of winter weather on the feeding habits of pheasants in southern Michigan. J. Wildl. Manage. 7:343–344.

Dillman, D. A. 1978. Mail and telephone surveys—the total design method. John Wiley & Sons, NY. 325pp.

Donnelly, D. M., and L. J. Nelson. 1986. Net economic value of deer hunting in Idaho. U.S.D.A. For. Serv. Res. Bull. RM–13. 27pp.

Emmerich, J. M. 1978. Bird utilization of woodland habitat in the eastern quarter of South Dakota. M.S. Thesis, South Dakota State Univ., Brookings. 130 pp.

————, and P. A. Vohs. 1982. Comparative use of four woodland habitats by birds. J. Wildl. Manage. 46:43–49.

Forman, R. T. T., and J. Baudry. 1984. Hedgerows and hedgerow networks in landscape ecology. Environ. Manage. 8:495–510.

Gates, J. E., and L. W. Gysel. 1978. Avian nest dispersion and fledging success in field-forest ecotones. Ecology 59:871–883.

Griffith, P. W. 1976. Introduction to the problems. Pages 3–7 *in* R. W. Tinus, ed. Shelterbelts on the Great Plains. Great Plains Agric. Council Publ. 78.

Hammack, J., and G. M. Brown, Jr. 1974. Waterfowl and wetlands: toward bioeconomic analysis. Johns Hopkins Univ. Press, Baltimore, MD. 95pp.

Hanson, W. R., and R. F. Labisky. 1964. Association of pheasants with vegetative types in east-central Illinois. Trans. N. Amer. Wildl. and Nat. Resour. Conf. 29:295–306.

Hintz, D. L., and J. R. Brandle, Editors. 1986. Proceedings of the international symposium on windbreak technology. Great Plains Agric. Council Publ. 117. 309pp.

Hudson P. J., and M. R. W. Rands, Editors. 1988. Ecology and management of gamebirds. BSP Professional Books, Oxford, UK 263pp.

Johnson, R. J., and M. M. Beck. 1988. Influences of shelterbelts on wildlife management and biology. Agric. Ecosystems & Envir. 22/23:301–335.

Kubantsev, B. S., and I. E. Vasil'ev. 1983. Composition, distribution, and numbers of birds in crop fields in northern regions along the lower Volga River. Soviet J. Ecol. 13:342–345.

LaPointe, D. F. 1958. Mourning dove production in a central Nebraska shelterbelt. J. Wildl. Manage. 22:439–440.

Lewis, T. 1969. The diversity of the insect fauna in a hedgerow and neighboring fields. J. Appl. Ecol. 6:453–458.

Love, D., and F. L. Knopf. 1978. The utilization of tree plantings by Mississippi Kites in Oklahoma and Kansas. Proc. For. Comm. Great Plains Agric. Council. 30:70–77.

_____, J. A. Grzybowski, and F. L. Knopf. 1985. Influence of various land uses on windbreak selection by nesting Mississippi kites. Wilson Bull. 97:561–565.

Lyon, L. J. 1954. Pheasant winter roosting cover preference in north-central Colorado. J. Wildl. Manage. 18:179–184.

_____. 1959. An evaluation of woody cover plantings as pheasant winter cover. Trans. N. Amer. Wildl. Conf. 24:277–289.

_____. 1961. Evaluation of the influences of woody cover on pheasant hunting success. J. Wildl. Manage. 25:421–428.

Martin, T. E. 1978. Diversity and density of shelterbelt bird communities. M.S. Thesis, South Dakota State Univ., Brookings. 174pp.

_____. 1981. Limitation in small habitat islands: chance or competition? Auk 98:715–734.

_____, and P. A. Vohs. 1978. Configuration of shelterbelts for optimum utilization by birds. Pages 79–88 *in* M. Craighead, R. Davis, R. Gardner and N. Smola, compilers. Trees—a valuable Great Plains multiple use resource. Great Plains Agric. Council Publ. 87.

May, J. F. 1978. Design and modification of windbreaks for better winter protection of pheasants. Ph.D. Thesis, Iowa State Univ., Ames. 175pp.

Norusis, M. J. 1988. SPSSPC+ V3.0. SPSS Inc., Chicago, IL. 117pp.

Petersen, L. 1979. Ecology of great horned owls and red-tailed hawks in southeastern Wisconsin. Dept. Nat. Resour. Tech. Bull. 111, Madison, WI. 63pp.

Podoll, E. B. 1979. Utilization of windbreaks by wildlife. Proc. For. Comm. Great Plains Agric. Council 31:121–128.

Popowski, J. 1976. Role of windbreaks for wildlife. Pages 110–111 *in* R. W. Tinus, ed. Shelterbelts on the Great Plains. Great Plains Agric. Council Publ. 78.

Potts, G. R. 1986. The partridge: pesticides, predation, and conservation. Collins, London, UK. 274pp.

Randall, R. N. 1955. Mourning dove production in south central North Dakota. J. Wildl. Manage. 19:157–159.

Read, R. A. 1958. The Great Plains shelterbelt in 1954. Great Plains Agric. Council Pub. 16. 125pp.

Rotzien, C. L. 1963. A cumulative report on winter bird population studies in eight deciduous shelterbelts of the Red River Valley, North Dakota. Proc. North Dakota Acad. Sci. 17:19–23.

Schaefer, P. R., S. Dronen, and D. Erickson. 1987. Windbreaks: a Plains legacy in decline. J. Soil and Water Cons. 42:237–238.

Schroeder, R. L. 1986. Habitat suitability index models: wildlife species richness in shelterbelts. U.S. Fish and Wildl. Serv. Biol. Rep. 82. 17pp.

Schroeder, W. R., and J. Kort. 1989. Shelterbelts in the Soviet Union. J. Soil and Water Cons. 44:130–134.

Schwilling, M. D. 1982. Nongame wildlife in windbreaks. Proc. For. Comm. Great Plains Agric. Council 34:258–262.

Shalaway, S. D. 1985. Fencerow management for nesting birds in Michigan. Wildl. Soc. Bull. 13:302–306.

Snyder, W. D. 1985. Survival of radio-marked hen ring-necked pheasants in Colorado. J. Wildl. Manage. 49:1044–1050.

Sorenson, C. J., and G. A. Marotz. 1977. Changes in shelterbelt mileage statistics over four decades in Kansas. J. Soil and Water Cons. 32:276–281.

Sorg, C. F., and J. B. Loomis. 1984. Empirical estimates of amenity forest values: a comparative review. Gen. Tech. Rep. RM-107, Rocky Mtn. For. and Range Exp. Stn., U.S. Dept. Agric., For. Serv., Ft. Collins, CO. 22pp.

Steinhoff, H. W., R. G. Walsh, T. J. Peterle, and J. M. Petulla. 1987. Evolution of the valuation of wildlife. Pages 34–48 *in* D. J. Decker and G. R. Goff, eds. Valuing wildlife—economic and social perspectives. Westview Press, Boulder, CO.

Stormer, F. A., and G. L. Valentine. 1981. Management of shelterbelts for wildlife. Proc. For. Comm. Great Plains Agric. Council 33:169–181.

United States Department of Interior, Fish and Wildlife Service. 1988. 1985 National Survey of Fishing, Hunting, and Wildlife Associated Recreation. U.S. Govern. Print. Office, Washington, DC. 167pp.

United States Water Resources Council. 1983. Economic and environmental principles for water and related resources implementation studies. U.S. Govern. Print. Office, Washington, DC. 137pp.

Walsh, R. G. 1986. Recreation economic decisions: comparing benefits and costs. Venture Publishing, Inc., State College, PA. 637pp.

Weiser, C., and D. J. Hlavini. 1956. Comparison of breeding-bird populations of three deciduous shelterbelts. Audubon Field Notes 10:416–419.

Wint, G. 1978. Utilization of shelterbelts by game species. Great Plains Agric. Council Pub. 87:93–95.

Yahner, R. H. 1981. Avian winter abundance patterns in farmstead shelterbelts: weather and temporal effects. J. Field Ornith. 52:50–56.

_____. 1982*a*. Avian nest densities and nest-site selection in farmstead shelterbelts. Wilson Bull. 94:156–175.

_____. 1982*b*. Avian use of vertical strata and plantings in farmstead shelterbelts. J. Wildl. Manage. 46:50–60.

_____. 1983*a*. Seasonal dynamics, habitat relationships, and management of avifauna in farmstead shelterbelts. J. Wildl. Manage. 47:85–104.

_____. 1983*b*. Small mammals in farmstead shelterbelts: habitat correlates of seasonal abundance and community structure. J. Wildl. Manage. 47:74–84.

© ED HOAG '90

3

Managed Habitats for Deer in Juniper Woodlands of West Texas

FRED C. BRYANT

EVERGREEN JUNIPER woodlands have increased gradually throughout central Texas since the 1870s. A major cause of this phenomenon is the complex interrelationship between reduced fire frequency and early settlement of the area, which included heavy grazing by domestic livestock. The landscape resulting from the livestock industry features vast tracts of grassland-juniper woodlands that are now regarded as valuable for wildlands, wildlife habitat, fuelwood supplies, scenic resources, and a potential outlet for expanding urban populations, as well as for livestock grazing.

Dr. Bryant points out the opportunities for sustained yields and multiple uses that these landscapes offer to the wise land planner. He further suggests the need for new approaches as development encroaches on juniper woodlands. Proper use of juniper woodlands can achieve practical results that meet the needs of rangeland managers and at the same time stimulate new initiatives in both conservation and multiple use. Such use requires that land planners view the regional ecosystem in holistic terms in which wildland, wildlife habitat, fuelwood, recreational, and development values can be integrated.

Bryant believes the present landscape will not be eradicated or totally replaced with another dominant land use. He therefore advocates conscious planning for the use of all resources in a manner where natural resources, economic values, and the diverse development of the landscape can be achieved.

INTRODUCTION

Over most landscapes where it occurs, juniper (*Juniperus* spp.) is the dominant vegetative feature. By virtue of its robust growth form and bright green hue, juniper is the most conspicuous plant even to the casual observer. This is particularly true in winter when all deciduous plants lose their leaves. For example, where juniper is codominant with the deciduous mesquite tree (*Prosopis glandulosa*), its perceptual dominance on the landscape is striking.

Whether in association with mesquite or live oak (*Quercus virginianus*), evergreen juniper is a codominant tree on millions of hectares of Texas's Edwards Plateau, Trans-Pecos, and Rolling Plains ecoregions. In east Texas, eastern red cedar (*J. virginiana*) occurs. Three primary species occur in west Texas. Ashe juniper (*J. ashei*) and one-seeded juniper (*J. monosperma*) occupy well over 8 million ha of west-central Texas. Redberry juniper (*J. pinchotii*), a stabilized hybrid of alligator (*J. deppeana*) and one-seeded (*J. monosperma*), is found on 10 million ha in western and central Texas. Four other species of juniper of lesser importance also occur in Texas (Gould 1975). Densities today may range from 200 to 4,000 plants per ha, with densities of 240 to 2,000 per ha commonly observed.

Because of its aesthetic appeal, one cannot overstate the importance of juniper and its place in juniper woodlands in any attempt to modify the environment or habitat. Real estate values in the famed Hill Country in Texas probably are linked to the presence, not absence, of juniper. When junipers are removed from the landscape, either a spindly mesquite or a rocky outcropping remains. The picture is not nearly as bleak where juniper is codominant with live oak, but if removed, juniper becomes conspicuously absent to the birder, hunter, land speculator, or urbanite.

The most abundant and popular native ungulate in Texas—the white-tailed deer (*Odocoileus virginianus*)—is more than casually associated with juniper and juniper woodlands. These economically and aesthetically important animals, although not restricted to juniper habitats, depend upon juniper in many ways. Moreover, in the western and northern counties of Texas, where redberry juniper extends its range, mule deer (*O. hemionus*) are found in association with this important tree. The role of juniper to the welfare of numerous other wildlife species, such as song-birds, lagomorphs, and furbearers, is an important consideration as well.

This chapter points out both positive and negative values of juniper and suggests management goals in juniper woodlands that consider: (1) biological needs of deer and other wildlife; (2) problems encountered by the livestock industry; and (3) the importance of juniper in an aesthetically managed landscape.

PAST, PRESENT, AND FUTURE LAND USES

Prior to 1850, junipers in Texas were restricted geographically and ecologically to steep canyon walls along drainages, buttes, escarpments, and steeper hillsides where fires were excluded. By contrast, fire maintained prairies relatively free of woody plants. Of these junipers, some were fire intolerant (*J. ashei* and *J. monosperma*), meaning fire severely damaged or killed the plant. Even though others resprouted readily after fire (*J. pinchotii* and *J. scopulorum*), fire frequency kept their woody biomass in check and hid their conspicuous growth behind tall grasses from early explorers and settlers. Early records from central Texas suggest that white-tailed deer also were associated primarily with drainages and steeper terrain.

With the rapid growth of cattle herds, and to some extent sheep and goats, the flourishing livestock industry from 1870 to the early 1920s allowed juniper to invade the prairies and grasslands; the greatest increase may have occurred between 1854 and 1884. Repeated grazing by livestock put relentless pressure on grasses, rendering them less able to compete with invading juniper for soil moisture and nutrients. Natural fires were suppressed because of public outcry and decreased fuel loads. Hence, the absence of fire invited an invasion by juniper and other woody plants on to grasslands. White-tailed deer herds expanded their range and grew in size concomitant with this increase and expansion of woody plant cover.

Second only to mesquite, juniper also was a successful invader on the Edwards Plateau and Rolling Plains. In fact, mesquite apparently facilitates initial juniper establishment on upland sites. Besides providing the microclimate or a "nurse tree" for establishing juniper seedlings, mesquite acted as a natural depository for seeds via avifauna consumption and subsequent dispersal. Thus, juniper may have followed mesquite on to the prairies.

So successful was juniper as an invader that ranchers began eradication and control measures probably as early as the 1930s and certainly by 1940. As juniper stands grew thicker and canopies closed, ranchers saw that juniper suppressed and eliminated herbaceous forage for livestock. Further, locating and handling livestock was difficult on juniper-dominated ranges. The necessity for frequent handling of livestock was particularly important during the years when the dreaded screwworm fly (*Cochliomyia hominivorax*) affected livestock. Juniper eradication measures included hand-cutting, mechanical treatments, and chemical application. But because the eradication measures between 1940–1970 rarely, if ever, integrated controlled fires—the natural suppressant—juniper always returned after two decades. Research on burning redberry juniper communities suggested canopy cover returns to preburn levels 13 years after burning. White-tailed deer numbers on the central portion of the Edwards Plateau exploded by 1950. The increase in brush, predator control, and control of the screwworm fly contributed to the rapid increase in deer numbers. Only the drought of the 1950s temporarily set back the population. From the 1960s through the early 1980s, white-tailed deer numbers were high enough for wildlife biologists to express their deep concern for habitat and animal quality.

Current trends reveal growing interests in real estate on the Edwards Plateau, as land values increase and ranchers intensify their management for white-tailed deer. Real estate values have become a complex mixture of aesthetic values associated with juniper woodlands and habitat values associated with the presence of white-tailed deer and other important wildlife species. Based on a model by Pope et al. (1984), an increase of one deer harvested per 259 ha correlates with an additional $45.24 per ha in the value of rural land. In 1981, with an approximate harvest of 10 deer per 259 ha in several counties in the Texas Hill Country, hunting contributed about $450 per ha to rural land values. Pope et al. (1984) specifically point out that this value probably partially reflects the value of other wildlife and the amenities of wildlife habitat. I propose that juniper is one of these amenities. Consequently, landowners today think less about juniper eradication and more about juniper management as urban development advances into the Hill Country.

On the juniper-dotted Rolling Plains, the numbers of deer hunters and hunter days have doubled since 1979, and the deer harvest has tripled. Because this ecoregion is far from major population centers, real estate values tied to aesthetics or deer habitat have remained relatively constant. However, hunting lease values have risen over the last 10 years.

In Texas, as the human population and recreational demands increase, more and more pressure will be placed on the Hill Country and the juniper woodland therein. The population of Texas is expected to double over the next 30 years; recreation in the form of hunting, hiking, birding, and camping will rise accordingly. The desire for solitude in an aesthetically pleasing outdoor setting will force, through economics, large tracts of land to be developed into smaller and smaller 4 to 40-ha tracts called *ranchettes* or *ranchitos*. Relative to juniper woodlands and deer habitat, this may result in a reinvasion of juniper onto level topography and/or fertile soils once managed as "juniper-free" by ranchers. Although other ranchette owners may decide to clear juniper, juniper probably will be allowed to consume the landscape as thickets. Relative to livestock, ranchettes will range from having too many livestock to none at all.

The impact of these alternatives on the integrity of white-tailed deer habitat is unknown. To be sure, the use of prescribed fire to control juniper will diminish. Unfortunately, only on large tracts of land (i.e., >200 ha) will any consideration be given to managed habitats for deer in juniper woodlands. Juniper woodlands in the Rolling Plains will be less affected, if at all. Juniper will spread and invade where it is adapted unless progressive ranchers and adjacent owners of ranchettes launch control measures. Even the most progressive rancher should consider juniper in managed landscapes for deer habitat management.

IMPORTANCE OF JUNIPER TO WILDLIFE

Deer

Even though juniper is considered an emergency food of low value, both white-tailed and mule deer use its leaves and berries at certain times of the year (Table 3.1.). On the Edwards Plateau, juniper comprised more than 20% of the diet of white-tailed deer in January and March (Waid et al. 1984). In their annual diet, deer consumed an average of 12% and contin-ued to use juniper even when other browse and forbs were available. Mule

TABLE 3.1.

The Contribution (%) of Juniper to the Diets of White-tailed and Mule Deer in Texas

Ecoregion, County	White-tailed Deer		Mule Deer		Reference
	Winter (%)	Annual (%)	Winter (%)	Annual (%)	
Edwards Plateau					
Kerr County[1]	37	30	—	—	Jackley (unpub. data)
Kerr County[2]	20	12	—	—	Waid et al. (1984)
Kerr County[2]	21	12	—	—	Jackley (unpub. data)
Edwards County[2]	24	9	—	—	Bryant (unpub. data)
Edwards County[3]	Trace	Trace	—	—	Bryant (1977)
Rolling Plains					
Oldham County	—	—	27	11	Sowell et al. (1985)
Donley County	—	—	15	6	Sowell et al. (1985)
Armstrong County	—	—	8	4	Krysl et al. (1979)
Trans-Pecos					
Culberson and Hudspeth counties	—	—	13	6	Krysl (1979)

[1] Poor range condition
[2] Fair range condition
[3] Excellent range condition; juniper was removed by root-plowing

deer in the Texas Panhandle use juniper in differing proportions depending upon the availability of oak (*Quercus* spp.). Where oak was absent or rare, mule deer consumed juniper at a rate of 6% to 11% annually, with highs of 15% to 27% in winter (Sowell et al. 1985). Where oak was abundant, mule deer favored oak over juniper, and dietary amounts of juniper declined to less than 4% annually; however, juniper still reached a high of 8% in winter (Krysl et al. 1979). In far-west Texas, mule deer ate at least 13% juniper in winter and 6% yearlong in the Guadalupe Mountains (Krysl 1979).

Juniper as cover for deer in Texas has not been studied adequately. But whether in the Edwards Plateau, Trans-Pecos, or Rolling Plains ecoregions, deer undoubtedly use juniper for cover. Thermal cover, described as that cover sought by animals to achieve energy balance, is provided by juniper in both winter and summer. During cold, windy periods, the dense leaf structure and robust growth afford deer shelter not provided by other shrubs. Wind velocity is reduced and temperatures are more stable and less extreme. For example, in Oregon, structural quality of a juniper stand created a microclimate in winter that was 40% less severe than that in an adjacent shrub land (Leckenby 1978). In another study where winters were severe, Bright (1978) reported that a juniper woodland adjacent to a shrub land had a microclimate that was 222% less severe, according to his cumulative weather index. That same dense foliage provides deer with shade on hot, summer days, although deer must seek junipers that do not block prevailing winds and gentle breezes. Hiding cover is used by deer to escape predators, to conceal daily movements, and to protect and conceal fawns. For mule deer in the Texas Panhandle, juniper breaks were preferred over all other vegetation types in winter and spring (Koerth et al. 1985). Although the specific role of juniper in habitat selection is unknown, the importance of this vegetation type for mule deer cannot be overemphasized. The extent that juniper provides security cover for white-tailed deer was demonstrated in central Texas by Rollins et al. (1988). Removal of more than 70% of standing, live juniper, even in well-designed patterns of small openings, resulted in lower deer use of clearings and lower numbers of deer (Fig. 3.1.).

Other Wildlife

Songbirds, furbearers, turkey (*Meleagris gallopavo intermedia*), and even threatened and endangered species depend on juniper at some time of the year. In Texas, little research has been done on the interrelationships of wildlife and juniper. By comparison, in Oregon, 83 species of birds and 23 species of mammals use western juniper (*J. occidentalis*) at some stage of the life cycle (Maser and Gashwiler 1978). Of these, 29 of the bird species were yearlong residents of the juniper woodland. Maser and Gashwiler (1978) also described structural changes in juniper as it matures, and discussed functional roles that offer advantages for numerous animals. For example, junipers can be used by birds for nesting, nest materials, perches, thermal cover, courtship, and food; and by small mammals as shade, shelter, nest materials, and food.

Furbearers, such as raccoons (*Procyon lotor*), skunks (*Spilogale* sp. and *Mephitis* sp.), gray foxes (*Urocyon cinereoargenteus*), ringtails (*Bassariscus astutus*), coyotes (*Canis latrans*), and bobcats (*Felis rufus*), eat juniper berries extensively during autumn and winter. Wild turkeys also relish juniper berries.

At least one endangered species is an obligatory member of juniper

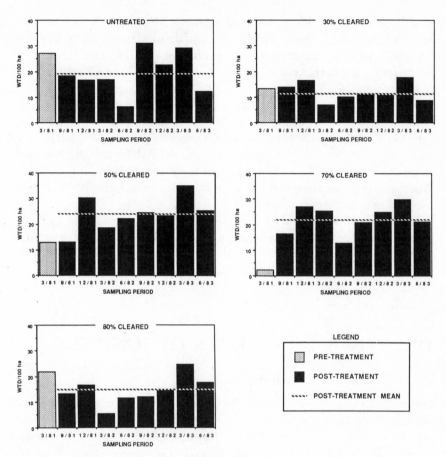

FIGURE 3.1.

Response of white-tailed deer to several intensities of juniper canopy removal on the Edwards Plateau in Texas (Rollins et al. 1988).

woodlands. The golden-cheeked warbler (*Dendroica chrysoparia*) requires overmature Ashe and redberry juniper stands for successful nesting. The females pull long strips of bark, obtained only from mature junipers, for their nests. According to Pulich (1976), the golden-cheeked warbler needs trees at least 50 years old. Ideal habitat would include trees 7 m in height, some adjacent deciduous oaks, with riparian area nearby to provide foraging perching sites for the birds after the nesting season. To support one breeding pair, 8 to 10 ha of dense juniper with intermingled oak trees would be required. Thus, recommendations for a breeding population include preserving a minimum of 125 ha of dense stands, and up to 2,000 ha where human encroachment is imminent. Unrestricted control of juniper may be detrimental to these birds, which have narrow habitat tolerances.

The extent to which the black-capped vireo (*Vireo atricapillus*), an endangered species, requires juniper is as yet undetermined. This vireo inhabits dry, limestone hilltops, ridges, and slopes on the eastern and southern portions of the Edwards Plateau. Its habitat usually contains oaks, mountain laurel (*Sophora secundiflora*), or sumacs (*Rhus* spp.). Often, Ashe juniper is associated with these shrubs. Sometimes juniper leaves are used for nest materials.

Last, juniper woodlands are important for winter residents, such as robins (*Turdus migratorius*) and cedar waxwings (*Bombycilla cedrorum*), that feed upon juniper mast. Large-scale removal of Ashe juniper may be detrimental to these species because of loss of winter food.

DELETERIOUS EFFECTS OF JUNIPER

Plant Community Development

Juniper is a vigorous competitor; it competes for moisture and nutrients with herbaceous flora. Its roots assume both horizontal and vertical positions in any soils they penetrate, and are capable of active absorption in both positions. Horizontal roots pose the more serious threat to herbaceous plants. Moreover, juniper is evergreen and transpiration occurs year-round, thus it uses soil moisture when most other plants are dormant. Extracts from its foliage have been shown to decrease grass growth and seed germination. These characteristics contribute to a more arid environment and create an erosion hazard. Shade-intolerant plants may be adversely affected, whereas shade-tolerant species may be enhanced.

The extent to which juniper suppresses grass and forb seedlings by virtue of its presence is also an important ecological consideration. Juniper has even been observed to invade well-established grasslands. For example, research indicates herbaceous cover may have little or no repressive effect on the establishment of juniper seedlings, but grass cover does slow growth rates and may act to delay flowering. McPherson (1987) determined that seedlings are established in wet years, particularly when fall and spring rains are adequate. In fact, more than 55% of the juniper seedlings in his study became established during the second year of a two-year wet cycle.

Juniper also may facilitate establishment of other shrubs, such as algerita (*Berberis trifoliolata*), littleleaf sumac (*Rhus microphylla*), and catclaw mimosa (*Mimosa biuncifera*) in the absence of grazing (McPherson 1987). As mentioned earlier, mesquite may facilitate establishment of redberry juniper seedlings. However, the density of mature juniper trees was unrelated to mesquite density; thus, this relationship is an age-dependent phenomenon because junipers were in only young age classes when related to mesquite.

The Livestock Industry

Wherever juniper woodlands occur, research has demonstrated the inverse relationship between juniper canopy cover and grass biomass and cover. Nearly all studies show logarithmic relationships. Closed juniper canopies exclude virtually all herbaceous vegetation.

In 1964, U.S.D.A. Soil Conservation Service reported that about 50% of the Edwards Plateau supported sufficient brush canopy to seriously suppress herbaceous forage production for livestock. Early work by Vallentine (1960) reported that areas cleared of Ashe juniper by hand cutting produced 6 times more forage than uncleared areas. But the response of herbaceous vegetation to overstory removal is site-specific; it depends upon soil depth, aspect, slope, and presence or absence of grazing. Rollins and Bryant (1986) studied the herbage response to chaining Ashe juniper followed in 60 days by slash cleanup with fire. Even though treated sites were generally on rocky areas with shallow soils, herbaceous biomass increased by 55% overall. Chaining reduced juniper canopy cover from 20% to 2%, and grass yield increased from 420 kg per ha to 660 kg per ha 22 months after treatment. Texas wintergrass (*Stipa leucotricha*), a cool-season grass important to livestock and deer, doubled on cleared sites. If sites are undisturbed except by livestock grazing for 3 to 4 years after chaining so that herbaceous plants can respond, burning increases grass yield by 50% in wet years (Wink and Wright 1976). Burning in dry years generally will reduce grass yields for several years.

Where redberry juniper is dominant in the Rolling Plains, grazing lends a competitive advantage to juniper (McPherson 1987). According to theories on competition between woody and herbaceous plants, the relationship is linear if there is no competition or logarithmic if competition exists. This relationship was linear on ungrazed sites and logarithmic on grazed sites. Furthermore, for every 5% increase in juniper canopy cover on ungrazed sites, grass yield declined by 4.3% (or 37 kg per ha). On grazed sites, the same 5% increase in juniper caused grass yield to decline by 26% (or 139 kg per ha). In general, a juniper canopy cover of 25% reduces grass yields by 50%.

An increase in livestock forage from juniper control results in greater grazing capacity. Moreover, improved animal performance may occur as forage plants with greater nutritional value become established. Because dense juniper impedes gathering of livestock, control measures that open up dense stands are helpful as well. Therefore, the economic benefits of juniper control to the livestock industry are obvious.

Wildlife Habitat

Extensive stands of dense juniper do not equate with optimum deer habitat. For example, dense juniper stands suppress desirable deer forage. On the Edwards Plateau, forb biomass was 5 to 6 times greater in spring and summer on a site cleared of juniper when compared to an adjacent closed stand of mature trees (Rollins and Bryant 1986). Important deer forages that increased were oaks (*Quercus* spp.), plaintains (*Plantago* spp.), and pellitory (*Parietaria pennsylvanica*).

Furthermore, dense stands become stagnant relative to other habitat values, such as plant diversity, structural contrast, and edge effect. Plant species richness, that is, number of species, was lower on uncleared sites relative to cleared sites (Rollins and Bryant 1986).

Our research on lagomorphs and songbirds illustrates some of the changes that can occur by clearing dense stands. Opening dense stands of Ashe juniper created a less-desirable habitat for cottontails (*Sylvilagus floridanus*) because they tend to prefer more closed habitats than do jackrabbits (*Lepus californicus*) (Fig. 3.2.). Cottontails generally decreased with clearing intensity except where 70% of the juniper canopy was removed (Rollins, 1983). This site also had deeper soils, and medium-to-tall grasses were more abundant. Because of shallower soils, shorter grasses such as common curly mesquite (*Hilaria belangeri*) were more abundant on the sites where 30%, 50%, and 80% of the juniper canopy was removed

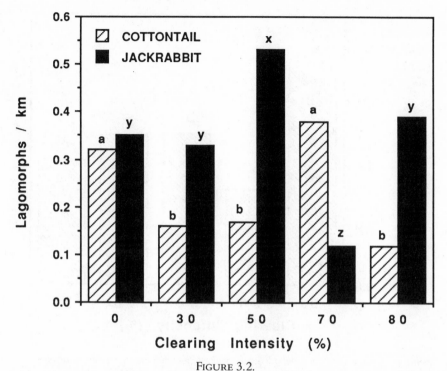

FIGURE 3.2.

Lagomorphs per km as determined from line-transect counts following five intensities of juniper removal, 1981–1982, in Kerr County, Texas. For each species, values with the same letters were not statistically different (P >0.05).

(Rollins and Bryant 1986). Thus, if tall grasses return after juniper control, cottontail numbers would be expected to return to relatively high numbers. If short grasses dominate after juniper clearing, jackrabbit abundance would be relatively greater. The grass community that prevails will be determined by both site and livestock-grazing strategies.

Clearing juniper at any intensity resulted in stable bird densities, but densities were lower than on the untreated site (Fig. 3.3.). Species diversity relative to untreated areas tended to increase as canopies were opened (Fig. 3.4.). Generally, when the landscape was changed from heavy amounts of juniper cover to more open habitats, tree-foraging birds were replaced by ground-foraging species (Fig. 3.5.; Rollins, 1983). Tree-foraging species included robins, common bushtits (*Psaltriparus minimus*), and black-crested titmice (*Parus bicolor*). Ground-foragers included spar-

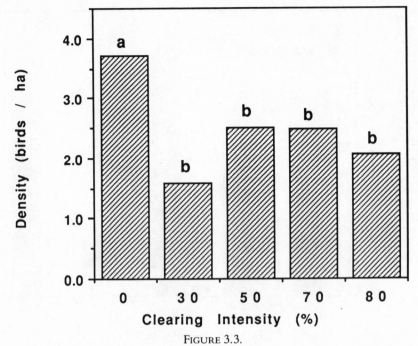

FIGURE 3.3.

Densities (birds per ha) of songbirds as determined from line-transect counts following five intensities of juniper removal, 1981–1892, in Kerr County, Texas. Values with the same letters were not statistically different (P > 0.05).

rows [lark (*Chondestes grammacus*), vesper (*Podecetes gramineus*), Lincoln's (*Melospiza lincolnii*) and rufous-crowned (*Aimophila ruficeps*)] and meadowlarks (*Sturnella* spp.). Also, American kestrels (*Falco sparverius*) and flycatchers (*Tyrannidae*) were common on cleared sites where they used juniper skeletons as perches (Rollins 1983). Brown towhees (*Pipilo fuscus*) and field sparrows were common near edges between cleared and untreated areas. Food availability no doubt increased with clearing. Greater biomass of annual forbs and large-seeded grasses was recorded on cleared areas (Rollins and Bryant 1986). For example, seeds of pellitory, which increased on cleared sites, are important foods for wintering Lincoln's sparrows. These lagomorph and avian responses were of short term (i.e., less than 2 years after clearing Ashe juniper). Secondary succession on cleared sites may continue for 10 years or more, with each seral stage favoring slightly different avian and microfaunal communities. Yet the relative benefits of opening up dense stands of juniper should be noted.

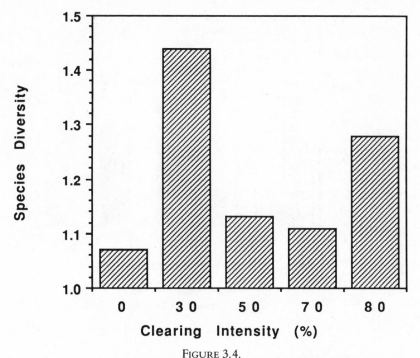

FIGURE 3.4.

Species diversity of songbirds following five intensities of juniper removal, 1981–1982, in Kerr County, Texas.

As discussed earlier, a closed canopy cover of juniper can exclude virtually all herbaceous vegetation. In the Rolling Plains, most forb species decrease as juniper cover increases. For example, croton (*Croton dioicus*), spurge (*Euphorbia lata*), scarlet gaura (*Gaura coccinea*), sensitive briar (*Shrankia uncinata*), and plains zinnia (*Zinnia grandiflora*) were negatively correlated with juniper cover (McPherson 1987). Wildlife using these plants should be expected to be enhanced if dense juniper stands are opened.

THE IMPORTANCE OF MANAGED LANDSCAPES

Managed landscapes in juniper woodlands of Texas suggest that people will make conscientious decisions in planning control measures for ju-

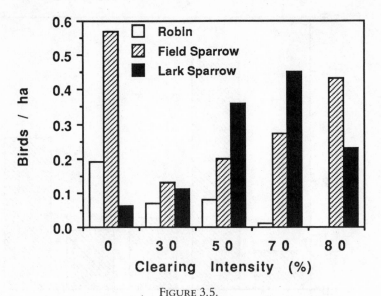

FIGURE 3.5.

Densities (birds per ha) of three songbirds following five intensities of juniper removal, 1981–1982, in Kerr County, Texas.

niper. These decisions must account for positive and negative implications concerning juniper:

- An area of closed, dense canopies that are generally undesirable implies that juniper will be controlled.
- Juniper's aesthetic appeal implies that control measures will include detailed landscape-planning techniques.
- Juniper's wildlife value implies that total eradication is undesirable.
- Regard for endangered and threatened species implies that careful consideration will be given to the area and sites where juniper is to be controlled.
- Knowledge of juniper ecology implies that, once created, openings will be maintained with controlled fire.
- Effect on future land values and/or deer habitat will dictate the strategy—amount and type of control measure—used in managing juniper woodlands.

Recommendations

EDWARDS PLATEAU

With the above implications in mind, the following are general recommendations for deer in managed habitats of juniper woodlands in west Texas.

Stands of sterile, tall (> 3 m) juniper exceeding 20% canopy cover should be opened in a landscape design. This design should include 8-ha clearings of irregular shape. Distance between clearings should be greater than 200 m to leave ample brush as security cover. Shape of openings should be generally rectangular with widths between 200 m to 350 m. Edges should be irregular rather than linear. No more than 50% of the total area should be cleared. If too little area is cleared, it could be as detrimental as too much. This is because spot grazing by all ungulates occurs if small openings are created in a sea of standing live juniper. However, clearings on steep slopes (greater than 15%) should be avoided; selective thinning is advised. Buffer strips 75 m to 125 m wide should be kept along roads and to shield openings made for powerlines; travel lanes 75 m to 125 m wide should be left for deer to access cultivated fields, food plots, fawning sites, and riparian areas. Slash and woody debris should not be removed entirely; perhaps 10% to 15% of a clearing should contain slash for use by other wildlife, such as songbirds and lagomorphs. Live oaks should be untouched. Any treatment of shin oak (*Q. pungens*) would depend upon its relative density. Some mature junipers should be kept in openings for aesthetic appeal as should narrow (10 m) corridors of trees. Prescribed fire scheduled at 7 years, 20 years, and 40 years posttreatment should be used to maintain these clearings. Livestock grazing should be controlled carefully. Floral diversity should be encouraged.

Large blocks of juniper that obviously impede livestock handling or suppress deer and livestock forage should be thinned in a landscape design. Thinning to 5% or 10% canopy cover of juniper can be accomplished by removing selected, preferably smaller (3 m) trees. Mosaics of juniper interspersed with oaks and other shrubs can create an aesthetic view. Creation of a savannahlike appearance tends to be more appealing visually, provided that adequate cover remains for deer. Such scenic impacts should be related to other goals and should be considered in the initial planning stages. Planners should identify distinctive types of landscape composition, contrasts of elements in color and line, and the variety of different parts of the landscape as they appear together. Planners should recognize access paths, such as roads and trails, and areas of high recreational importance as being prime management zones for aesthetics. Refer to Schreyer and Royer (1975) for more detail.

Where trees are still small (< 3 m), prescribed burning may be used to open up these sites. Mosaics of burned and unburned areas are most desirable.

ROLLING PLAINS AND TRANS-PECOS

Juniper is much less a target for control in the Rolling Plains and Trans-Pecos ecoregions. Dense junipers on rough, steep slopes should be selectively thinned or left alone for optimum deer management. These sites provide hiding and thermal cover. On more level to rolling topography dense juniper may warrant opening these stands to provide deer with foraging areas. The description of punching holes in a sea of juniper best applies, which can be done following the recommendations for the Edwards Plateau. Clearings or selective thinning is recommended because habitats take on an "applied landscaping" look.

Literature Cited

Bright, L. 1978. Weather stress difference between two levels of juniper canopy cover. Pages 91–95 in R. E. Martin, J. E. Dealy, and D. L. Caraher, eds. Proc. Western Juniper Ecol. and Manage. Workshop. U.S.D.A. For. Serv. Gen. Tech. Rep. PNW–74, Bend, OR.

Bryant, F. C. 1977. Botanical and nutritive content in diets of sheep, Angora goats, Spanish goats, and deer grazing a common pasture. Ph.D. Thesis, Texas A&M Univ., College Station. 92 pp.

Gould, F. W. 1975. Texas plants—a checklist and ecological summary. MP-585 Tex. A&M Univ. and Tex. Agric. Exp. Stn., College Station.

Koerth, B. H., B. F. Sowell, F. C. Bryant, and E. P. Wiggers. 1985. Habitat relations of mule deer in the Texas panhandle. Southwest. Nat. 30:579–587.

Krysl, L. J. 1979. Food habits of mule deer and elk, and their impact on vegetation in Guadalupe Mountains National Park. M.S. Thesis, Texas Tech Univ., Lubbock. 131 pp.

————, C. D. Simpson, and G. G. Gray. 1980. Dietary overlap of sympatric Barbary sheep and mule deer in Palo Duro Canyon, Texas. Pages 97–103 in C. D. Simpson, ed. Proc. Symp. on Ecol. and Manage. of Barbary Sheep. Dep. Range and Wildl. Manage. Texas Tech Univ., Lubbock.

Leckenby, D. A. 1978. Western juniper management for mule deer. Pages 137–161 in R. E. Martin, J. E. Dealy, and D. L. Caraher, eds. Proc. Western Juniper Ecol. and Manage. Workshop. U.S.D.A. For. Serv. Gen. Tech. Report PNW–74, Bend, OR.

Maser, C., and J. S. Gashwiler. 1978. Interrelationships of wildlife and western juniper. Pages 37–82 *in* R. E. Martin, J. E. Dealy, and D. L. Caraher, eds. Proc. Western Juniper Ecol. and Manage. Workshop. U.S.D.A. For. Serv. Tech. Rep. PNW-74, Bend, OR.

McPherson, G. R. 1987. Plant interactions in redberry juniper-mixed grass communities. Ph.D. Thesis, Texas Tech Univ., Lubbock. 87 pp.

Pope, C. A. III, C. E. Adams, and J. K. Thomas. 1984. The recreational and aesthetic value of wildlife in Texas. J. Leisure Res. 16:51–60.

Pulich, W. M. 1976. The golden-cheeked warbler. Texas Parks and Wildl. Dept., Austin. 172 pp.

Rollins, D. 1983. Wildlife response to different intensities of brush removal on the Edwards Plateau of Texas. Ph.D. Thesis, Texas Tech. Univ., Lubbock. 70 pp.

Rollins, D., and F. C. Bryant. 1986. Floral changes following mechanical brush removal in central Texas. J. Range Manage. 39:237–240.

Rollins, D., F. C. Bryant, D. D. Waid, and L. C. Bradley. 1988. Deer response to brush management in central Texas. Wild. Soc. Bull. 16:277–284.

Schreyer, R., and L. E. Royer. 1975. Impacts of pinyon-juniper manipulation on recreation and aesthetics. Pages 143–151 *in* G. F. Gifford and F. E. Busby, eds. Proc. The Pinyon-Juniper Ecosystem: A Symposium. Utah State Univ., Logan.

Sowell, B. F., B. H. Koerth, and F. C. Bryant. 1985. Seasonal nutrient estimates of mule deer diets in the Texas Panhandle. J. Range Manage. 38:163–167.

Vallentine, J. F. 1960. Live oak and shin oak as desirable plants on Edwards Plateau ranges. Ecology 41:545–548.

Waid, D. D., R. J. Warren, and D. Rollins. 1984. Seasonal deer diets in central Texas and their response to brush control. Southwest. Nat. 29:301–307.

Wink, R. L., and H. A. Wright. 1976. Effects of fire on an Ashe juniper community. J. Range Manage. 26:326–329.

©ED HOAG

4

Browse Diversity and Physiological Status of White-tailed Deer During Winter

GLENN D. DELGIUDICE, L. DAVID MECH,
AND ULYSSES S. SEAL

THE IDEA of hidden relationships and common bonds existing between plant and animal life has always had an appeal to wildlife-habitat researchers. The primary reason for this appeal is based upon the knowledge that wildlife persists only under favorable combinations of habitat conditions.

A basic way to ascertain the health and dynamics of wildlife populations is to determine a carrying capacity of their specific habitat. Traditional approaches have focused on the physical properties of vegetative cover and wildlife's utilization of that vegetation. However, the patterns and details of each species's utilization of vegetation still left our understanding of wildlife-habitat interrelationships unanswered. More recently, wildlife biologists have turned to a chemical assessment of blood and urine to determine the nutritional status of the animal. This type of analysis exposes patterns of use and relationships between the nutritional content of plants and physiological status of animals that previously were not visible.

Drs. DelGiudice, Mech, and Seal have examined the blood and urine chemistry of certain wildlife species in order to determine better the relationship between animals and the habitat in which they reside. Their findings suggest a complex relationship between the dynamics of plant nutrition, forage availability, and habitat conditions, and the animals changing physiology that allows them to adapt to these changes. Research, it has been stated, is not supposed to tell us why things are the way they are in the natural world, but rather it strives to increase our knowledge about these phenomena. And among the field of wildlife researchers are those who feel compelled to put their ideas of these phenomena to some use. Drs. DelGiudice, Mech, and Seal direct their research to a point where people can, with the proper information, manage for a higher degree of predictability regarding wildlife survival over winter.

INTRODUCTION

Estimation of carrying capacity is the principal means by which biologists and managers relate the health and dynamics of deer (*Odocoileus* spp.) populations to the quality of their changing habitat (Mautz 1978, Harlow 1984, McCullough 1984). Although many habitat factors influence carrying capacity, ecologists concur that nutrition ultimately has the most direct effect on health, size and productivity of deer herds (Verme and Ullrey 1972, Mautz 1978, Robbins 1983). Management decisions related to the carrying capacity for deer are often based on evaluations of vegetational conditions; however, the value of this information is limited by the accuracy of the assessment and may be misleading (Van Horne 1983), especially when used as an indirect measure of deer condition or nutritional status. A more refined understanding of the relationship between browse quantity and quality and the physiological status of deer is needed for more informed decisions related to carrying capacity (Halls 1984).

The obvious and direct consequences of gross-habitat deterioration and poor nutrition on deer reproduction and survival have been well-documented (Morton and Cheatum 1946, Cheatum and Severinghaus 1950, Julander et al. 1961, Verme 1965, 1969, Smith and LeCount 1979). However, recent evidence has indicated that there are subtle and indirect nutritional influences as well (Nelson and Mech 1986, Mech et al. 1987, O'Gara and Harris 1988).

Thorough examination and understanding of these influences and of the relationship between habitat (i.e., food resources) and deer condition requires the ability to detect and quantify subtle temporal and spatial differences in deer nutrition (DelGiudice et al. 1989). Urinalysis as a direct, sensitive means of assessing the physiological status of deer (Warren et al. 1981, 1982, DelGiudice et al. 1987, 1988a) combined with analyses of snow-urine (urine deposited in snow), a technique that permits sequential and frequent sampling of free-ranging deer (DelGiudice et al. 1988b, 1989), should help fulfill this requirement. Our objective in this

79

study was to examine the relationship between browse availability and use in winter yards of white-tailed deer (*O. virginianus*) and the physiological status of deer in those yards from early to late winter.

STUDY AREA

Three winter yards located within the east-central Superior National Forest (SNF) in northeastern Minnesota (48°N, 92°W) (DelGiudice et al. 1989) were included in our study. The Kawishiwi (KAW), Snort Lake (SL) and Isabella (ISA) yards occupied 7, 6, and 27 km², respectively (Nelson and Mech 1987). Kawishiwi and SL each held 40 to 50 deer: about 400 deer concentrated in the ISA yard (Nelson and Mech 1986, 1987).

Northeastern Minnesota is characterized by undulant topography and a continental climate (Anon. 1978). Approximately 127 cm of snow fell from mid-November 1984 to early April 1985 (National Oceanic and Atmospheric Administration 1984, 1985). DelGiudice et al. (1989) provide additional weather information for the study area.

Mixed coniferous-deciduous stands were prominent on the uplands and included balsam fir (*Abies balsamea*), white spruce (*Picea glauca*), paper birch (*Betula papyrifera*), trembling aspen (*Populus tremuloides*), jack pine (*Pinus banksiana*), and northern white cedar (*Thuja occidentalis*). Beaked hazel (*Corylus cornuta*), mountain maple (*Acer spicatum*), red-osier dogwood (*Cornus stolonifera*), and speckled alder (*Alnus rugosa*) are shrubs apparent on the site (Wetzel et al. 1975). Conifer swamps were associated with the lowlands, populated by black spruce (*Picea mariana*), tamarack (*Larix laricinia*), northern white cedar, bog birch (*Betula pumila*), and Labrador tea (*Ledum groenlandicum*). Logging of hardwoods occurred on site for pulp production and to enhance habitat quality for deer (T. R. Biebighauser, U. S. For. Serv., pers. comm.).

METHODS

We compared food habits of deer in the 3 yards during winter 1984–85 to determine if there was a relationship between percent browse availability and use and nutritional status as assessed by chemical analysis of snow-urine. We employed a technique for examining food habits similar to that used earlier in the same region (Wetzel et al. 1975). During early (15 January–15 February) and late (16 February–15 March) winter, we located fresh deer tracks in each yard after a new snowfall. Availability and fresh browsing of 6 "indicator" species (mountain maple, red-osier dogwood, trembling aspen, beaked hazel, paper birch, and balsam fir) and a seventh category called "other" were documented within 1 m of both sides of the tracks for about 450 m. We chose the 6 indicator species because of their varied levels of importance to deer (Wetzel et al. 1975, Rogers et al. 1981). A freshly browsed twig was recorded as an "instance of use" (Wetzel et al. 1975). Each twig of current year's growth was enumerated as available browse. Significantly ($P < 0.05$) greater percent use compared to percent availability indicated preference for a species; less use suggested avoidance (Wetzel et al. 1975).

Deer were tracked and food habits recorded 14 to 23 times in each yard during both early and late winter. We divided a map of each yard into 10 grid cells, and sampling was distributed over the cells. Percent browse availability and use data were subjected to arc-sine transformation before comparisons were made by analysis of variance. Duncan's multiple range test was employed to make group comparisons at $P < 0.05$.

We collected snow-urine samples after recent snowfalls during 5 2-week intervals: (1) 1–15 January; (2) 16–31 January; (3) 1–15 February; (4) 1–15 March; and (5) 16–31 March (DelGiudice et al. 1989). Collections began at ISA during Interval 2. Snow-urine samples were assayed for urea nitrogen (U), sodium (Na), potassium (K), phosphorus (P), and creatinine (C). Urea N and electrolyte data were compared as ratios to C to control extraneous variability associated with single urine samples and to correct for dilution by snow. Details of the snow-urine collections and chemical analyses are described elsewhere (DelGiudice et al. 1989).

RESULTS

Browse Availability and Use

In the general study area, mean proportional availability of trembling aspen declined from early (6.3 ± 1.2% [SE]) to late (3.4 ± 0.9%) winter ($P < 0.05$). Similarly, red-osier dogwood availability tended ($P = 0.08$) to diminish (2.7 ± 0.7% versus 1.3 ± 0.5%); percent use also decreased

($P < 0.05$) during late winter (10.6 ± 2.8% versus 4.7 ± 1.5%). Availability of balsam fir was greater ($P < 0.05$) in late winter (16.0 ± 1.8%) than during early winter (11.5 ± 1.8%). Mountain maple was highly preferred throughout winter, and red-osier dogwood was preferred during early winter and selected according to availability in late winter.

During early and late winter, significant ($P < 0.001$) differences in proportional browse availability and selection occurred among the 7 browse categories within each of the 3 yards (Tables 4.1. and 4.2.). At KAW, mean availability of browse classified as "other" (15 spp.) was greater than that of all other categories during early and late winter (Tables 4.1. and 4.2.). Although mountain maple was more available than only a few other indicator species throughout winter (Tables 4.1. and 4.2.), its selection by deer was similar to that of "other" and greater than that of all remaining indicator species (Tables 4.1. and 4.2.). Use of "other" (8–10 spp.) was only greater than that of a few other indicator species during early winter (Table 4.1.), but during late winter it comprised a greater portion of the deer's diet than all other indicator species except mountain maple (Table 4.2.). All remaining indicator species were selected in proportion to availability throughout winter.

Similar to KAW, browse classified as "other" was most available (19 spp.) at SL during early and late winter (Tables 4.1. and 4.2.). However, "other" (11 spp.) was browsed more frequently than any other indicator species throughout winter as well. Selection of "other" included the following: willows (*Salix* spp.), speckled alder, pin cherry (*Prunus pennsylvanica*), choke cherry (*P. virginiana*), honeysuckle (*Lonicera* spp.), blueberry (*Vaccinium* sp.), northern white cedar, black ash (*Fraxinus nigra*), bog birch, and two unidentified species. Mountain maple was browsed more frequently than each of the 5 remaining indicator species in early winter, even though there were no differences in availability (Table 4.1.). Although greater use of mountain maple than of most species continued during late winter, selection of beaked hazel was similar at this time (Table 4.2.).

In contrast to KAW and SL, beaked hazel at ISA was most available in early winter (Table 4.1.). During the same period, availability of "other" (15 spp.) was greater than that of most other indicator species. By late winter, availabilities of "other" (13 spp.) and beaked hazed were similar, and each was more available than each of the other indicator species (Table 4.2.). During early winter, beaked hazel and "other" (8 spp.) were selected by deer more than any other indicator species except mountain maple (Table 4.1.); however, mean proportional use of beaked hazel by late

TABLE 4.1.

Proportional Availability and Use (%) of Twigs of 7 Indicator Species of Browse by White-tailed Deer in 3 Yards in Northeastern Minnesota, Early Winter (15 Jan–15 Feb), 1985

Indicator browse species	Kawishiwi Availability x̄	SE	Use x̄	SE	Snort Lake Availability x̄	SE	Use x̄	SE	Isabella Availability x̄	SE	Use x̄	SE
Trails n	18		18		21		21		15		15	
Mountain maple	21.8A[a]	6.3	35.2A	9.2	8.0A	2.7	22.1A	6.2	5.1AB	2.8	20.3AB	7.9
Red-osier dogwood	4.0BC	1.8	14.5BC	6.6	2.3A	1.0	7.2B	3.1	1.5B	0.7	7.5BC	4.6
Trembling aspen	8.2AB	2.0	2.0C	0.7	5.8A	2.3	0.9B	0.5	4.4AB	1.5	4.5C	4.5
Beaked hazel	15.3AC[b]	4.6	14.4BC[c]	6.7	8.6A	2.4	4.9B	2.7	46.7C	7.6	30.4A	8.2
Paper birch	1.4B	0.5	2.9C	1.1	3.6A	1.8	7.5B	3.2	4.9AB	1.7	7.0BC	3.9
Balsam fir	14.8AC	4.1	0.0C	0.0	7.9A	2.6	0.5B	0.5	12.0AD	2.4	0.0C	0.0
Other[d]	34.8D[e]	7.5	25.5AB[e]	8.6	63.7B	8.2	47.4C	9.4	25.4D	6.9	23.3A	5.9
Total	**100.3**		**94.5[f]**		**99.9**		**90.5[f]**		**100.0**		**93.0[f]**	

[a] Mean values in a column with different letters are different ($P < 0.05$).
[b] Significant difference ($P < 0.05$) between Isabella and Kawishiwi and Snort Lake.
[c] Significant difference ($P < 0.05$) between Isabella and Snort Lake.
[d] Includes an additional 13 to 17 and 8 to 10 species available and selected, respectively.
[e] Significant difference ($P < 0.05$) between Snort Lake and Kawishiwi and Isabella.
[f] The remaining portion is attributable to no selection by deer of available browse.

winter was greater than that of all other categories, including "other" and mountain maple (Table 4.2.).

Balsam fir was avoided with respect to consumption in all 3 yards throughout winter. Trembling aspen was avoided during early winter and browsed according to availability during late winter at KAW and SL. However, at ISA it was browsed in proportion to availability throughout winter.

WINTER YARD COMPARISONS

Mountain maple, red-osier dogwood, trembling aspen, beaked hazel, paper birch, and balsam fir accounted collectively for 66% and 72–75% of available browse in the KAW and ISA yards, respectively, throughout winter (Tables 4.1. and 4.2.). These species represented only 36% and 52% of browse available to deer at SL during early and late winter. Throughout winter, beaked hazel was more available at ISA than at KAW and SL (Tables 4.1. and 4.2.). Availability of browse classified as "other" was greater at SL than at KAW during early winter and greater than at ISA throughout winter (Tables 4.1. and 4.2.).

Mountain maple, red-osier dogwood, trembling aspen, beaked hazel, paper birch, and balsam fir collectively constituted 66% to 69% and 70% to 78% of browse selected by deer at KAW and ISA (Tables 4.1. and 4.2.). During early winter, just 2 species, mountain maple and beaked hazel, together represented almost two-thirds and three-quarters of this use at KAW and ISA, respectively (Table 4.1.). By late winter, this portion of use by deer increased to ≥ three-quarters in both yards (Table 4.2.).

All winter, mean selection of beaked hazel at ISA was greater than at SL (Tables 4.1. and 4.2.). During late winter, use of beaked hazel at ISA was greater than at KAW as well (Table 4.2.). Only 23% and 22% of total browse used by deer at ISA was attributable to "other" during early and late winter. At KAW, "other" was associated with 26% and 34% of total browse use during these periods. However, use of the numerous browse species of this category by SL deer during early winter (47%) was greater than by deer at KAW and ISA (Table 4.1.). During late winter, mean use of "other" at SL (53%) remained greater than at ISA, but was similar to its use at KAW (Table 4.2.).

Snow-urine Profiles

Urea nitrogen:C declined steadily in snow-urines at KAW ($P < 0.0001$) and ISA ($P < 0.001$) throughout winter; whereas U:C remained stable

TABLE 4.2.

Proportional Availability and Use (%) of Twigs of 7 Indicator Species of Browse by White-tailed Deer in 3 Yards in Northeastern Minnesota, Late Winter (16 Feb–15 Mar), 1985

Indicator browse species	Kawishiwi				Snort Lake				Isabella			
	Availability		Use		Availability		Use		Availability		Use	
	x̄	SE	x̄	SE	x̄	SE	x̄	SE	x̄	SE	x̄	SE
Trails n	21		21		19		19		22		22	
Mountain maple	21.0A[a]	6.3	35.7A	9.0	12.3AB	4.4	21.9A	6.9	12.6	4.0	23.9	5.9
Red-osier dogwood	0.7B	0.6	2.7BC	1.6	1.2A	0.7	4.1B	2.5	1.9	1.2	6.0	3.4
Trembling aspen	4.3B	1.8	9.6BC	4.2	5.5AB	2.2	3.8B	2.3	0.9	0.3	0.6	0.5
Beaked hazel	18.2A[b]	4.9	15.9C[b]	5.7	17.6B	5.5	13.0AB	5.1	39.3	5.8	42.5	6.8
Paper birch	0.9B	0.4	1.9BC	0.9	3.3A	1.7	3.7B	1.6	3.2	1.3	4.7	2.1
Balsam fir	20.5A	3.8	0.0B	0.0	12.6B	2.8	0.7B	0.7	14.5	2.8	0.7	0.5
Other[c]	34.4C[d]	4.5	34.2A[d]	8.0	47.6C	8.7	52.9C	9.5	27.9	5.4	21.6	6.7
Total	100.3		100.0		100.1		100.1		100.3		100.0	

[a] Mean values in a column with different letters are different ($P < 0.05$).

[b] Significant ($P < 0.05$) difference between Isabella and Kawishiwi and Snort Lake.

[c] Includes an additional 12 to 14 and 8 to 11 species available and selected, respectively.

[d] Significant ($P < 0.05$) difference between Snort Lake and Isabella.

through early March at SL, then declined ($P < 0.05$) during late March (Fig. 4.1.) (DelGiudice et al. 1989). Mean U:C was greater ($P < 0.05$) in snow-urines at SL than in samples at KAW and ISA during early March (DelGiudice et al. 1989).

At KAW, snow-urine Na:C ($P < 0.005$) and K:C ($P < 0.15$) decreased progressively until late March when both ratios increased significantly (P 0.05) (Fig. 4.1.) (DelGiudice et al. 1989). Sodium:C ($P < 0.01$) and K:C ($P < 0.05$) at ISA also decreased progressively, but remained diminished through late March (Fig. 4.1.) (DelGiudice et al. 1989). At SL, Na:C remained stable through late March; K:C was unaltered through early March but became elevated ($P < 0.05$) during late March (Fig. 4.1.) (DelGiudice et al. 1989).

At KAW and SL, mean P:C did not change throughout most of winter, but P:C increased ($P < 0.05$) by late March in both yards (Fig. 4.1.) (DelGiudice et al. 1989). Snow-urine P:C at ISA increased ($P < 0.01$) slightly until late March when it declined ($P < 0.05$) (Fig. 4.1.).

DISCUSSION

Generally, frequent browsing of mountain maple and beaked hazel by deer, and preferences for mountain maple and red-osier dogwood in the study area, were similar to findings from previous studies (Wambaugh 1973, Wetzel et al. 1975). Although percent availability and use of most browse species did not change as winter progressed, increasing snow depth appeared to influence the decline in availability of already scarce red-osier dogwood and trembling aspen, which inhabit primarily open areas where snow accumulation is greatest. Similarly, diminished browsing of red-osier dogwood coincided with a period when snow covered many of the low-growing dogwoods, and deer used coniferous cover heavily (Wetzel et al. 1975). Low use of trembling aspen was consistent with observations during past winters (Wambaugh 1973, Wetzel et al. 1975, Mooty 1976) and may be related to its limited nutritive quality (Ullrey et al. 1971).

Complete avoidance by deer of balsam fir for food throughout winter suggested that nutritional inadequacy was not severe enough for deer to select this lower quality browse (Ullrey et al. 1968, Klein 1970). Furthermore, rebrowsing of previously selected current year's growth was not observed (DelGiudice, unpubl. data). Rogers et al. (1981) noted that consumption of balsam fir by starved deer (Aldous and Smith 1938) during late winter was much greater than by deer considered to be in good condition (Wetzel 1972).

Sequential collection and chemical analysis of snow-urine directly and physiologically confirmed that deer in all 3 yards remained in an early phase of undernutrition throughout winter (DelGiudice and Seal 1988, DelGiudice et al. 1989). Low levels of U:C (< 4.0) and electrolyte ratios with C (Fig. 4.1.) conformed with documented patterns of reduced home-range size and decreased movement and feeding by deer as snow accumulates (Rongstad and Tester 1969, Ozoga and Verme 1970, Moen 1978). These ratios also indicated that fat reserves of deer were not exhausted and that extensive endogenous protein catabolism was not occurring (Waid and Warren 1984, Torbit et al. 1985, DelGiudice and Seal 1988, DelGiudice et al. 1989).

Early undernutrition was associated with diverse diets in deer of all 3 yards as suggested by the varied use of the 6 indicator species and the continued importance of numerous species that constituted the seventh category of "other." However, there were differences among yards in browse availability, dietary diversity, and the nutritional status of deer (Fig. 4.1.) (DelGiudice et al. 1989).

Steady declines of U:C, Na:C, and K:C in snow-urines of KAW and ISA deer from early and late January, respectively, to early March (Fig. 4.1.) indicated progressive nutritional deprivation (DelGiudice et al. 1989) and reflected physiological mechanisms of nutrient conservation (Robbins et al. 1974, Robbins 1983). The diet of SL deer was more diverse, and unaltered mean levels of U:C, Na:C, K:C, and P:C until early March in these deer (Fig. 4.1.) suggested that they were able to maintain a more constant level of nutritional adequacy (DelGiudice et al. 1989). By early March, when snow was deepest, greater U:C ratios in SL deer than in KAW and ISA deer indicated that dietary protein availability was greater in the former (DelGiudice et al. 1989). Crude protein and mineral contents vary widely among browse species in the general area (Peek et al. 1976, DelGiudice 1988), but locational influences within species and season were likely minor (Short et al. 1966).

Similar to domestic ruminants, deer selection of plant species appears to be related to content of specific nutrients (Swift 1948, Weir and Torell 1959). Dietary diversity probably provides more adequate levels of various nutrients by overcoming specific mineral deficiencies that limit

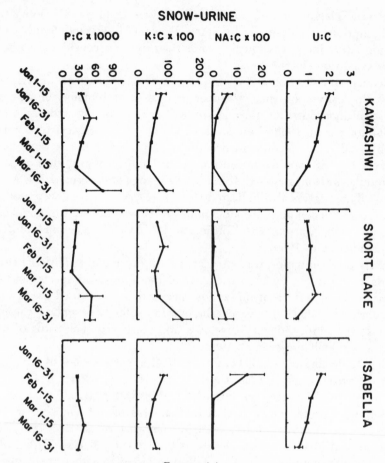

FIGURE 4.1.

Mean (± SE) ratios of urea nitrogen (U), sodium (Na), potassium (K), and phosphorus (P) to creatinine (C) in white-tailed deer urine in snow (snow-urine) collected during five two-week intervals in the Kawishiwi, Snort Lake, and Isabella yards, northeastern Minnesota, 1985 (compiled from DelGiudice et al. 1989).

digestion of some plants (Church 1977, Hanson and Jones 1977). Furthermore, evidence has shown that deer maintain body weights better on more heterogeneous diets during winter, and even plants consumed in small quantities are probably nutritionally beneficial (Dahlberg and Guettinger 1956, Verme and Ullrey 1972).

Dietary diversity of SL deer was evidenced by their greater selection of the 11 additional browse species of "other" compared to each of the remaining 6 indicator species and by the major portion of their diet that "other" comprised throughout winter (Tables 4.1. and 4.2.). Although "other" species were consistently present at KAW and ISA, sparser availability at KAW during early winter, and at ISA throughout winter, compared to SL indicated less variety of available browse. Similarly, the lower proportional use of this browse category reflected less variety in the diets of KAW deer in early winter and ISA deer during the entire winter compared to the diets of SL deer.

The more limited dietary diversity of KAW and ISA deer also was reflected by the observation that just 2 species (mountain maple and beaked hazel) constituted most of their diet. This was particularly evident at ISA, where consistently greater browsing of beaked hazel and less selection of "other" species compared to SL deer suggested dramatic differences in diet diversity throughout winter. Deeper snow at ISA (DelGiudice et al. 1989) and the association of beaked hazel with dense, coniferous cover probably contributed to the deer's frequent selection of beaked hazel as a forage item (Wetzel et al. 1975). This was indicated further by the change from similar proportional selection by ISA deer of beaked hazel, mountain maple, and "other" during early winter to the predominance of beaked hazel in their diet during late winter when snow was deepest (DelGiudice et al. 1989).

Browse composition on site, plant availability, and snow depth are the primary factors affecting the winter diet of free-ranging ruminants (LeResche and Davis 1973). An additional indicator of the more diminished nutrition available at ISA was the deer's selection of trembling aspen according to availability throughout winter (Klein 1970, Ullrey et al. 1971). Deer at KAW and SL avoided trembling aspen during early winter and used this species in proportion to availability in late winter.

During late winter, the similar availability and consumption of "other" by KAW and SL deer was indicative of a relative increase in dietary diversity in KAW deer, even though mountain maple still constituted a major portion of their diet. Although greater U:C ratios during early March (Fig. 4.1.) suggested greater nutritional intake in SL deer than KAW deer, the apparent increase in dietary diversity in KAW deer during late winter may have contributed to initiation of a similar nutritional recovery during late March (DelGiudice et al. 1989). The late winter recovery in SL and KAW deer was indicated by significant ($P < 0.05$) elevations of Na:C, K:C, and P:C during late March (Fig. 4.1.) and was associated with minimal snow cover (< 18 cm) at both yards (DelGiudice et al. 1989), as well as elevated metabolic demands and feeding activity (Silver et al. 1969, Ozoga and Verme 1970, DelGiudice et al. 1987).

Increased nitrogen retention and accretion efficiency, a concomitant with the improved dietary protein supply subsequent to protein deficiency, were probably responsible for the lack of simultaneous elevations of U:C ratios in KAW and SL deer (Fig. 4.1.) (Deb.Hovell et al. 1987, DelGiudice et al. 1989). Absence of a similar nutritional recovery in ISA deer by late March, indicated by C ratios of U, Na, K, and P that remained diminished, seemed attributable to prolonged deep snow (≥ 35 cm) and less dietary diversity.

MANAGEMENT IMPLICATIONS

We have shown by simultaneous study of deer physiology (via snow–urine analysis) and browse availability and selection in 3 yards that subtle differences in available food resources and diet diversity may be accompanied by quantifiable differences or changes in the metabolic status of free-ranging deer as winter progresses. The ability to sequentially detect and quantify such disparities is an important step toward a more accurate understanding of the ecological relationship between deer condition and habitat quality and what constitutes optimum deer habitat. Our data also reflected the obvious potential of snow cover to affect browse availability and thus, deer nutritional status. Although certain environmental vagaries are beyond the control of managers, our data strongly suggest that maximizing browse diversity should be a key consideration in management plans for deer habitat improvement. The seemingly subtle nutritional effects reported herein may have important implications related to deer reproduction and survival, and dictate further study.

Acknowledgments

We appreciate financial support from the following: Special Projects Foundation, Minneapolis Big Game Club; National Rifle Association; U.S. Fish and Wildlife Service; U.S.D.A. North Central Forest Experiment Station; Research Service, Veterans Administration Medical Center; Minnesota Zoological Society; R. Shelton; Safari Club International; James Ford Bell Foundation; and Minnesota Department of Natural Resources. Statistical consultation was provided by F. Martin, and early drafts of the manuscript were reviewed by S. E. Jorgensen and Y. Cohen. We thank the following for technical assistance: M. D. Lewis, M. K. Twite, N. I. Manning, J. W. Burch, N. L. Barton, L. L. Laack, N. J. Dietz, and D. H. Monson.

Literature Cited

Aldous, S. E., and C. F. Smith. 1938. Food habits of Minnesota deer as determined by stomach analysis. Trans. N. Amer. Wildl. Conf. 3:756–767.

Anonymous. 1978. Soil survey of Kawishiwi area, Minnesota, parts of Lake and Cook counties in Superior National Forest. U.S. Dep. Agric., Soil Cons. Serv. and Minn. Agric. Exp. Stn. 36pp.

Cheatum, E. L., and C. W. Severinghaus. 1950. Variations in fertility of white-tailed deer related to range conditions. Trans. N. Amer. Wildl. Conf. 15:170–189.

Church, D. C. 1977. Livestock feeds and feeding. D. C. Church, publisher, Corvallis, OR. 349pp.

Dahlberg, B. L., and R. C. Guettinger. 1956. The white-tailed in Wisconsin Tech. Wildl. Bull. 14. Wisconsin Conserv. Dep., Madison. 282pp.

Deb.Hovell. F. D., E. R. Orskov, D. J. Kyle, and N. A. MaCleod. 1987. Undernutrition in sheep. Nitrogen repletion by N-depleted sheep. Br. J. Nutr. 57:77–88.

DelGiudice, G. D. 1988. Physiological assessment of winter nutritional status of white-tailed deer (*Odocoileus virginianus*) in Minnesota by urine and blood analyses. Ph.D. Thesis. Univ. Minnesota, St. Paul. 136pp.

DelGiudice, G. D., L. D. Mech, and U. S. Seal. 1988*b*. Chemical analyses of deer bladder urine and urine collected from snow. Wildl. Soc. Bull. 16:324–326.

————. 1989. Physiological assessment of deer populations by chemical analysis of urine in snow. J. Wildl. Manage. 53. 284–291.

DelGiudice, G. D., L. D. Mech, U. S. Seal, and P. D. Karns. 1987. Winter fasting and refeeding effects on urine characteristics in white-tailed deer. J. Wildl. Manage. 51:860–864.

DelGiudice, G. D., and U. S. Seal. 1988. Classification of winter undernutrition in white-tailed deer via serum and urinary urea nitrogen. Wildl. Soc. Bull. 16:27–32.

DelGiudice, G. D., U. S. Seal, and L. D. Mech. 1988*a*. Response of urinary hydroxyproline to dietary protein and fasting in white-tailed deer. J. Wildl. Dis. 24:75–79.

Halls, L. K. 1984. Research problems and needs. Pages 783–790 *in* L. K. Halls, ed., White-tailed deer: ecology and management. Stackpole Books, Harrisburg, PA.

Hanson, H. C., and R. L. Jones. 1977. The biogeochemistry of blue, snow, and Ross' geese. Ill. Nat. Hist. Survey, Urbana. 320pp.

Harlow, R. F. 1984. Habitat evaluation. Pages 601–628 *in* L. K. Halls, ed., White-tailed deer: ecology and management. Stackpole Books, Harrisburg, PA.

Julander, O., W. E. Robinette, and D. A. Jones. 1961. Relation of summer range condition to mule deer herd productivity. J. Wildl. Manage. 25:54–60.

Klein, D. R. 1970. Food selection by North American deer and their response to over-utilization of preferred plant species. Pages 25–44 in A. Watson, ed., Animal populations in relation to their food resources. Blackwell Sci. Publ. Ltd., Oxford, UK.

LeResche, R. E., and J. L. Davis. 1973. Importance of nonbrowse foods to moose on the Kenai Peninsula, Alaska. J. Wildl. Manage. 37:279–287.

Mautz, W. W. 1978. Nutrition and carrying capacity. Pages 321–348 in J. L. Schmidt and D. L. Gilbert. eds., Big game of North America: ecology and management. Stackpole Books, Harrisburg, PA.

McCullough, D. R. 1984. Lessons from the George Reserve, Michigan. Pages 211–242 in L. K. Halls, ed., White-tailed deer: ecology and management. Stackpole Books, Harrisburg, PA.

Mech, L. D., R. E. McRoberts, R. O. Peterson, and R. E. Page. 1987. Relationship of deer and moose populations to previous winters' snow. J. Anim. Ecol. 56:615–627.

Moen, A. N. 1978. Seasonal changes in heart rates, activity, metabolism, and forage intake of white-tailed deer. J. Wildl. Manage. 42:715–738.

Mooty, J. J. 1976. Year-round food habits of white-tailed deer in northern Minnesota. Minnesota Wildl. Res. Quart. 36:11–36.

Morton, G. H., and E. L. Cheatum. 1946. Regional differences in breeding potential of white-tailed deer in New York. J. Wildl. Manage. 10:242–248.

National Oceanic and Atmospheric Administration. 1984. Climatological data: Minnesota. Nat. Climatic Cent., Asheville, NC. 355pp.

————. 1985. Climatological data: Minnesota. Nat. Climatic Cent., Asheville, NC. 371pp.

Nelson, N. E., and L. D. Mech. 1986. Deer population in the Central Superior National Forest, 1967–1985. U.S.D.A. For. Serv. Res. Pap. NC–271. 8pp.

————. 1987. Demes within a northeastern Minnesota deer population. Pages 27–40 in B. G. Chepko-Sade and Z. Halpin, eds., Mammalian dispersal patterns. Univ. Chicago Press, Chicago, IL.

O Gara, B. W., and R. B. Harris, 1988. Age and condition of deer killed by predators and automobiles. J. Wildl. Manage. 52:316–320.

Ozoga, J. J., and L. J. Verme. 1970. Winter feeding patterns of penned white-tailed deer. J. Wildl. Manage. 34:341–349.

Peek, J. M., D. L. Urich, and R. J. Mackie. 1976. Moose habitat selection and relationships to forest management in northeastern Minnesota. Wildl. Monogr. 48. The Wildlife Society. Washington, DC. 65pp.

Robbins, C. I. 1983. Wildlife feeding and nutrition. Academic Press, NY. 343pp.

————, R. L. Prior, A. N. Moen, and W. J. Visek. 1974. Nitrogen metabolism of white-tailed deer. J. Anim. Sci. 38:871–876.

Rogers, L. L., J. J. Mooty, and D. Dawson. 1981. Foods of white-tailed deer in the Upper Great Lakes region—a review. U.S.D.A. For. Serv. Gen. Tech. Rep. NC–65. 24pp.

Rongstad, O. J., and J. R. Tester. 1969. Movements and habitat use of white-tailed deer in Minnesota. J. Wildl. Manage. 33:366–379.

Short, H. L., D. R. Dietz, and E. E. Remmenga. 1966. Selected nutrients in mule deer browse plants. Ecology 47:222–229.

Silver H., N. F. Colovos, J. B. Holter, and H. H. Hayes. 1969. Fasting metabolism of white-tailed deer. J. Wildl. Manage. 33:490–498.

Smith, R. H., and A. LeCount. 1979. Some factors affecting survival of desert mule deer fawns. J. Wildl. Manage. 43:657–665.

Swift, R. W. 1948. Deer select most nutritious forages. J. Wildl. Manage. 12:109–110.

Torbit, S. C., L. H. Carpenter, D. M. Swift, and A. W. Aldredge. 1985. Differential loss of fat and protein by mule deer during winter. J. Wildl. Manage. 49: 80–85.

Ullrey, D. L., W. G. Youatt, L. D. Fay, B. E. Brent, and K. E. Kemp. 1968. Digestibility of cedar and balsam fir browse for the white-tailed deer. J. Wildl. Manage. 32:162–171.

Ullrey, D. L., W. G. Youatt, H. E. Johnson, L. D. Fay, D. B. Purser, and B. L. Schoepke. 1971. Limitations of winter aspen browse for the white-tailed deer. J. Wildl. Manage. 35:732–742.

Van Horne, B. 1983. Density as a misleading indicator of habitat quality. J. Wildl. Manage. 47:893–901.

Verme, L. J. 1965. Reproduction studies on penned white-tailed deer. J. Wildl. Manage. 29:74–79.

————.1969. Reproductive patterns of white-tailed deer related to nutritional plane. J. Wildl. Manage. 33:881–887.

————, and D. L. Ullrey. 1972. Feeding and nutrition of deer. Pages 275–291 *in* D. C. Church, ed. Digestive physiology and nutrition of ruminants. Vol. 3, Practical nutrition of ruminants. Oregon State Univ., Corvallis.

Waid, D. D., and R. J. Warren. 1984. Seasonal variations in physiological indices of adult female white-tailed deer in Texas. J. Wildl. Dis. 20:212–219.

Wambaugh, J. R. 1973. Food habits and habitat selection by white-tailed deer and forage selection in important habitats in northeastern Minnesota. M.S. Thesis, Univ. Minnesota, St. Paul. 71pp.

Warren, R. J., R. L. Kirkpatrick, A. Oelschlaeger, P. F. Scanlon, and F. C. Gwazdauskas. 1981. Dietary and seasonal influences on nutritional indices of adult male white-tailed deer. J. Wildl. Manage. 45:926–936.

Warren, R. J., R. L. Kirkpatrick, A. Oelschlaeger, P. F. Scanlon, K. E. Webb, Jr., and J. B. Whelan. 1982. Energy, protein, and seasonal influences on white-tailed deer fawn nutritional indices. J. Wildl. Manage. 46:302–312.

Weir, W. C., and D. T. Torell. 1959. Selective grazing by sheep as shown by a comparison of the chemical composition of range and pasture forage obtained by hand clipping and that collected by esophageal fistulated sheep. Anim. Sci. 18:641–649.

Wetzel, J. F. 1972. Winter food habits and habitat preferences of deer in northeastern Minnesota. M. S. Thesis, Univ. Minnesota, St. Paul. 106pp.

————, J. R. Wambaugh, and J. M. Peck. 1975. Appraisal of white-tailed deer winter habitats in northeastern Minnesota. J. Wildl. Manage. 39:59–66.

© ED HOAG '90

5

Conservation of Rain Forests in Southeast Alaska: Report of a Working Group

FRED B. SAMSON, PAUL ALABACK, JERE CHRISTNER,
THOMAS DeMEO, ARLENE DOYLE, JON MARTIN,
JAMES McKIBBEN, MARK ORME, LOWELL SURING,
KENNETH THOMPSON, BRUCE G. WILSON,
DAVID A. ANDERSON, RODNEY W. FLYNN,
JOHN W. SCHOEN, LANA G. SHEA, and
JERRY L. FRANKLIN

THE SOUTHEAST Alaskan rain forest represents one of the most unique ecosystems found on the North American continent. Perhaps more unique than the ecosystem is the land-management problem the region presents to the land planner. This area represents one of the best examples of saving an ecosystem by managing it as a regional resource rather than treating it as a unique environmental entity.

The Tongass National Forest, which rests wholly within a rain forest ecosystem, is the single largest national forest in North America. Under federal law, the U.S.D.A. Forest Service is obligated to prepare management plans for its use. Currently, the Forest Service is updating land-management plans completed in 1979 and again in 1982. To coordinate such an effort requires a regional-scale planning framework.

The issues include ecological, economic, social demographic, and ethical values. All these concerns are related to the efficient management of the forest's mineral, wildlife, water, and wilderness resources. Dr. Samson and his coauthors have put into a concisely stated perspective of the strategies needed to prioritize, protect, and harvest these resources in a prudent and timely manner.

Relatively speaking, these forests are intact and connected. Therefore, their wildlife habitats and wildlife have not yet suffered the impacts of decades of clear-cutting and poor management practices. If, however, the problems of habitat fragmentation are to be avoided, a more sophisticated plan for utilizing the resources must be proposed.

Samson recommends the protection of widely distributed key biological units by physiographic province so as to defray effects to the region caused by habitat fragmentation. To do this will require a full understanding of the various plant and animal communities operating with this unique environment. Furthermore, a creative approach to the harvesting of the forest resource within these provinces is needed to minimize the impacts of lumbering.

INTRODUCTION

Rain forests are among the most unique and limited ecosystems world-wide (Alaback 1988). The rain forest in North America, principally along the coast of southeast Alaska, is unique with Sitka spruce 200 feet tall, 400+ years old, with a lush undergrowth of evergreen plants, ferns, and mosses. Wildlife is abundant and unique, ranging in size from the Alaskan brown bear (*Ursus arctos*) to the Sitka black-tailed deer (*Odocoileus hemionus sitkensis*) to Peal's peregrine (*Falco peregrinus pealei*) to the Glacier Bay water shrew (*Sorex alaskanus*).

Most of the rain forest in southeast Alaska is part of the Tongass National Forest and managed for multiple use by the Forest Service (U.S.D.A. Forest Service 1988). Timber produced on the Tongass National Forest supports local economies and contributes to economies of Pacific Rim nations. Recreation and commercial fisheries are significant industries and, as with timber, impact regional economies and those of other nations. Furthermore, the unique blend of forest, wildlife and fisheries is significant to native cultures—Tlingit and Haida—both in tradition and as a source of subsistence, as well as to other Alaskans who live a subsistence lifestyle.

The need for a new way of thinking in conservation of natural resources has been suggested by many authors (Harris 1984, Cairns 1986, Noss 1987, Bourgeron 1988), but examples are few. These authors suggest that success in maintaining biological diversity—perhaps the most important resource on public lands (Wilcove 1988)—increases when the focus of conservation efforts is at the landscape level in contrast to an emphasis on a species, population, or individual (Noss 1983). The purpose of this paper is to summarize concepts developed by an interagency, interdisciplinary working group for the long-term management of North America's rain forest, given the recent landscape emphasis in conservation of biological diversity.

The rain forest in southeast Alaska extends north to south 500 miles, is about 100 miles wide, and is a mosaic of small to large offshore islands, deep fjords, and mainland, all with differing plant assemblages that extend from shoreline to well above tree line. Some, but not all, plant communities are intensively managed for timber harvest, a land use that does affect size of a plant community and, in some cases, composition.

A workshop was held 25 and 26 May, 1988, at the Juneau Ranger District, Tongass National Forest, to increase our understanding about the role of old-growth rain forest habitat and how best to manage old-growth to maximize habitat value to associated species. The workshop was based on a recommendation from the 1988 U.S.D.A. Forest Service Alaska Region Biologists' Conference to address specific management questions associated with developing and implementing an old-growth wildlife management prescription for the Tongass National Forest.

The approach taken by the working group was one of scale. First, define ecological units in southeast Alaska, specifically the old-growth rain forest plant communities. Second, establish a province system that captures representative samples of plant communities and thereby habitat for all species dependent on or closely associated with that old-growth habitat. Third, recommend size, shape and distribution of habitats in a way to increase the likelihood that viable populations of old-growth associated species will be maintained on the Tongass National Forest—a legal mandate on all national forests. In addition, the future for biological diversity on public lands rests not only with preserving representative samples of pristine ecosystems, but also with innovative management of intensively used landscapes. The third set of recommendations is to enhance biological diversity on intensively managed lands.

OLD-GROWTH HABITATS

Southeast Alaska is an area of coastal mainland and islands isolated from Canada and other regions of Alaska by high mountain ranges. These lands also include the largest remaining reserve of old-growth forest in the United States. Interest in southeast Alaska old-growth forests and wildlife-fishery resources associated with these habitats has increased in the last decade (Schoen et al. 1981, Schoen and Kirchoff 1988). Although recent scientific literature has increased our ability to define general characteristics of late successional forests, working definitions specific to southeast Alaska are needed to guide current planning efforts and to clarify issues in management of fish and wildlife resources dependent on late successional forests.

At least 3 concepts need to be considered in the development of an ecological definition of old-growth in southeast Alaska. First, the ecological definition should be community specific. Data available for the Alaska Region include plant association and timber inventory information developed by the Forest Service. Second, the definition should be multifaceted and include criteria related to the presence of large and/or old trees, intermediate-sized trees that contribute to a deep multilayered canopy, a coarse woody debris—particularly snags and down logs—and a varied species composition. Such ecological characteristics are important to distinguish clearly the characteristics of old-growth from either the early seedling-sapling or mature, even-aged forest successional stages. Third, area is a key element in an ecological definition. Sufficient stand size is important to preserve interior forest dynamics, maintain microclimate associated with old-growth stands, and ensure the long-term survival of old-growth stands.

In southeast Alaska, at least seven forest series exhibit a late successional component. They include three highly productive Sitka spruce-western hemlock associations (upland, riparian, and beach), a moderately productive Sitka spruce-western hemlock, mixed conifer, and subalpine-mountain hemlock. The basic criteria or minimum standards for each of the seven associations are summarized in Table 5.1.

OLD-GROWTH DISTRIBUTION

The landscape of southeast Alaska is a mosaic of heterogeneous landforms, vegetation types, and offshore islands that vary in size and shape. Most animal species that occupy that landscape are neither threatened nor endangered, range from abundant and widespread to uncommon and localized, and may be found in only a portion of southeast Alaska. Managing for viable populations and biological diversity in southeast Alaska must consider this landscape mosaic, extent and distribution of old-growth as influenced by timber management, and islands.

To meet diversity and viable population requirements as outlined in the Regulations to the National Forest Management Act (1976), the workshop

TABLE 5.1.

Description of seven old-growth associations found in Southeast Alaska. They are the highly productive Sitka spruce-western hemlock (HPU), highly productive Sitka spruce-western hemlock riparian (HPR), highly productive Sitka spruce-beach fringe (HPB), moderately productive western hemlock (MO), cedar-western hemlock (CWH), mixed conifer (MC), and subalpine mountain hemlock (SMH). Height is in feet, diameter in inches, age in years, area in acres, and all values are equal or greater than. Minimum area includes a core area and surrounding tree buffer of three tree heights. The buffer may be forest types other than core type with a tree height at least 75% of core height.

Tree/Stand Characterists	HPU	HPR	HPB	MO	CWH	MC	SMH
Tree height	120	130	130	80	80	60	45
Tree diameter	25	30	30	25	15	13	12
Tree age	200	200	200	200	200	200	200
Multilayered canopy	yes	yes	yes	yes	yes	yes	yes
Discontinuous canopy	yes	yes	yes	yes	yes	yes	yes
Snag height	12	30	20	15	15		15
Snag diameter	25	30	30	15	15		10
Snag number	2	2	2	8	2	15	8
Woody debre length	5	50	50	20	20		
Woody debre diameter	25	30	30	10	15		30
Woody debre number	4	4	4	8	4		4
Minimum area[a]	120		60	100	60	50	60
Minimum core size	60			50	35	30	35
Minimum core width	900		500	800	700	700	750
Minimum no. old-growth trees/acre	90–110	90–110	90–110	70–89	70–89	40–69	<40

[a] Minimum riparian area is a core area as wide as the riparian corridor. The buffer is the natural adjacent plan community. Core area is ½ mile in length. Minimum beach area is the width of the beach fringe zone 500′ from mean high tide. Length is based on requirements of bald eagles in southeast Alaska and is slightly more than 1 mile. A 1-mile by 500-foot zone results in a minimum area of 60 acres.

identified 18 provinces within southeast Alaska to account for differences in latitude, altitude, maritime versus terrestrial climates, and other geographic factors that affect the distribution of old-growth habitat types (Fig. 5.1.). An additional degree of resolution may be needed to meet viability requirements for species with limited distributions, and units of land averaging about 100,000 ac have been proposed for the revision of the Tongass Land Management Plan. This broad-scale geographic approach to viable populations and biological diversity is similar to recommenda-

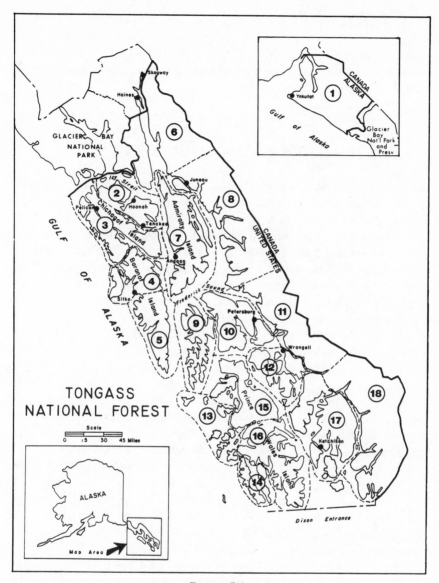

FIGURE 5.1.

Eighteen provinces are in southeast Alaska. They are (1) Yakutat, (2) Eastern Chichagof, (3) Western Chichagof, (4) Northern Baranof, (5) Southern Baranof, (6) Northern Lynn Canal, (7) Admiralty, (8) Taku-Endicott, (9) Kuiu, (10) Kupreanof, (11) Sitkinc, (12) Zarembo-Etolin-Wrangell, (13) Coronation-Heceta-Suemez Islands, (14) Dall-Sukkwan-Long Islands, (15) North Prince of Wales, (16) South Prince of Wales, (17) Reillagigedo-Island-Cleveland Peninsula, and (18) Misty Fiords.

tions of Urban et al. (1987) and Scott et al. (1987) and is employed in successful conservation programs by the Nature Conservancy (Rousch 1985).

MANAGING FOR OLD-GROWTH

How to develop alternatives for size, shape, and distribution of old-growth within a project area, often a watershed, is a task frequently encountered by biologists and other resource managers (Mealey et al. 1982, Harris 1984, Franklin and Forman 1987). During the past decade, a number of authors have raised the issue that patches of habitat need to be large enough to maintain breeding populations of wildlife dependent upon that habitat type. Among examples used to emphasize the need for large habitat patches are the eastern wood warbler complex (Wilcove 1985), resident and migratory grassland birds (Samson 1980, Risser 1986), grizzly bear (*Ursus arctos*) (Shaffer and Samson 1985), and northern spotted owl (*Strix occindentalis*) (U.S.D.A. Forest Service 1986). Furthermore, there is a need to emphasize a circular shape in managing habitat patches (Temple 1983) and to consider corridors that connect distant habitat patches (Harris 1984). The circular shape of habitat patches is thought to reduce negative effects of edge on some species by minimizing reduction of patch size resulting from windthrow. Corridors are viewed as important to allow for dispersal of individuals between habitat patches when individual patches are too small to support viable populations of some species.

The approach taken in the workshop was to offer general guidelines for managing old-growth rain forests within a watershed. Underlying each guideline is an emphasis on large, continuous blocks of old-growth forest needed to maintain viable populations. An increasing number of species-specific recommendations for patch size is available based on habitat suitability models developed for the Tongass Land Management Plan. In addition, no consideration was given to other land-use requirements, including the visual quality of the landscape, protecting anadromous fish resources, and the economics/techniques of timber harvest. The importance of these needs in forestland management was recognized during the workshop.

The three guidelines offered in this paper address management prescriptions that emphasize: (1) timber, (2) timber-wildlife, and (3) wildlife. In addition, five recommendations are offered to enhance diversity through time in harvest areas.

Timber Emphasis

Important in the timber emphasis is the harvest of timber in vertical, wide continuous strips (Fig. 5.2.). Size of old-growth patch to be harvested should be equal to or exceed in size a functional old-growth stand as outlined under old-growth series definitions. This will allow regeneration of functional, persistent old-growth stands given a sufficiently long timber rotation. The upper portion of the watershed should be harvested first. This will retain important lowland habitats during most of the first rotation. This approach will maximize the availability at any point in time of remaining lowland old-growth. Lowland old-growth habitats are particularly important to Sitka black-tailed deer, brown bears, bald eagles (*Haliaeetus leucocephalus*), and other wildlife.

In addition, the north aspect should be harvested in large vertical adjoining units before entering the south-facing slope. Where possible, the south-facing slope should be maintained as a contiguous unit of old-growth habitat types. South-facing slopes provide critical late winter habitat needed by wildlife, particularly the Sitka black-tailed deer. Maintaining large blocks of old-growth from the riparian and or shoreline up to treeline will provide a complete array of habitats and allow for seasonal movements between lowland and upland habitats.

Timber-Wildlife Emphasis

The extent of timber harvest would be less in areas with this management emphasis than where timber is the emphasis. Otherwise, the same guidelines apply. Although information to establish minimum old-growth forest patch size for species of wildlife in southeast Alaska is becoming available, a conservative approach requires that old-growth stands be maintained as large, continuous blocks connected by suitable travel corridors along riparian and beach fringe (Fig. 5.3.). This will increase the likelihood that the stands will be large enough for species that have not been adequately studied and stands will persist through time due to reduced extent of edge exposed to wind. Such an approach will also maximize at any point in time the amount of contiguous old-growth

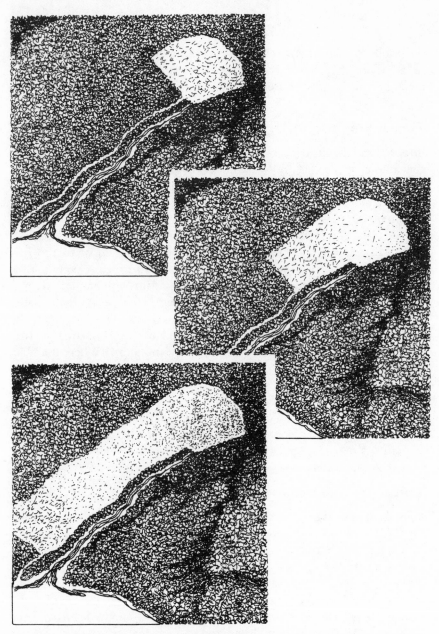

FIGURE 5.2.
Recommended pattern of harvest to emphasize timber production. Harvest begins in the upper portion of the watershed on north-facing slopes and proceeds toward the lower portion of the watershed.

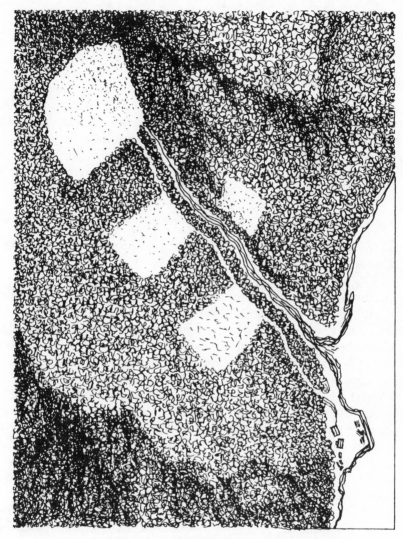

FIGURE 5.3.
Recommended pattern of timber harvest to emphasize timber-wildlife production. Harvest units are large and extend from riparian areas to treeline.

habitat for wildlife. Most of the time, blocks of old-growth will be mosaics of a combination of several old-growth types and may include naturally occurring muskegs.

Wildlife Emphasis—No Entry

The assumption under the wildlife emphasis is that continuous, undisturbed old-growth provides maximum benefit to those species dependent upon or closely associated with old-growth. Maintenance of entire watersheds of old-growth forest would maintain viable populations for most, if not all, species. The group recommends that at least one watershed within each province be left intact.

Harvest and Biological Diversity

Five additional guidelines recommended by the workshop that relate to the timber and timber-wildlife emphases are:

1. Harvest areas should be large and continuous. Harvest of old-growth should proceed from the periphery inward. The locus method illustrated in Figure 5.4. leaves at any point in time the largest contiguous block of old-growth within any cutting unit. This approach is important to provide the maximum amount of old-growth and to minimize the amount of edge habitat vulnerable to windthrow.
2. Harvest areas should be "sloppy" and include small patches of green trees, brushy openings, and snags. Leaving green trees will provide, through time, needed snag habitat within often monotypic second-growth stands. Likewise, brushy openings provide forage for species such as the Sitka black-tailed deer for which typical second-growth stands have limited value. The purpose is to increase through time the diversity of habitats available to wildlife in managed forests.
3. Edges of harvest units should be "feathered" rather than sharp. This reduces vulnerability (in contrast to a distinct edge) of forest stands to windthrow—a significant source for loss of old-growth timber resources.
4. Old-growth habitat types should be harvested to ensure the continued existence of each old-growth forest type and the relative availability of each type.
5. Habitat models for management indicator species identified for the Tongass Land Management Plan revision should be used to prioritize which areas should be retained for old-growth forest wildlife habitat.

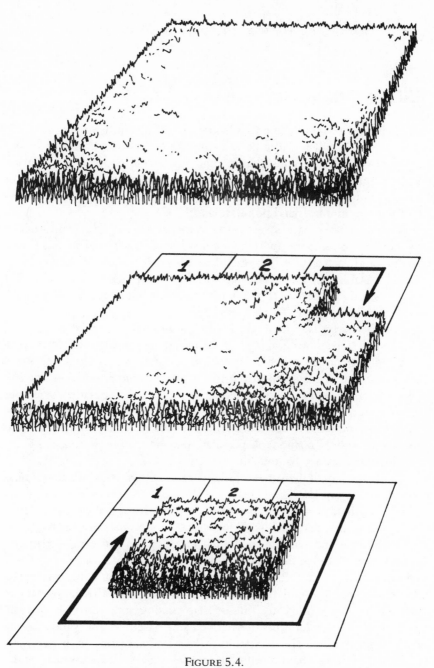

FIGURE 5.4.
A forest stand illustrating the locust method of timber harvest. The locust method retains the most old-growth at any point in time and reduces extent of edge habitats.

ISLANDS: A SPECIAL CASE

The complex of mainland and offshore islands in southeast Alaska varies in size from small to large and in distance to mainland. Rarely in any archipelago are all islands able to maintain viable populations of large, wide-ranging species. Some islands may not even maintain populations of small, common species. In almost every island system, dispersal is important in maintaining island populations.

Whether a given barrier serves to reduce species dispersal from 1 island to another is dependent upon the life histories of individual species (MacArthur and Wilson 1967), which are poorly understood in southeast Alaska. Managers, however, must consider how many, what size, and which islands are needed to maintain well-distributed and viable populations of wildlife species in southeast Alaska.

One alternative is to maintain all old-growth habitats on islands up to a given size. The minimum size of an island to consider would be that required to maintain a viable population of a species thought to represent island species. A major approach to estimate viable population size has been to calculate Ne, the net effective breeding population size. Ne provides for short-term genetic diversity and therefore viability. Generally, 50 reproductively active individuals, half of which are females, are considered a short-term viable population using the Ne approach; 500 are considered necessary for long-term viability.

A common small mammal on islands in southeast Alaska is the ermine (*Mustela erminea*). Possible genetic isolation of this species on islands has resulted in the development of a unique subspecies. This subspecies has a home-range size that is thought to encompass home-range characteristics of other island wildlife for which viability is a concern. The multiplication of 50 times the average home-range size of an ermine (40 ac) suggests that at least 2,000 ac of forest habitat would be needed to maintain a viable population. The literature suggests that home ranges may be as large as 84 ac, but overlap among individuals is evident (Lockie 1966). Under this alternative, and based on the validation of ermine home range and habitat use (Fay 1985), islands with less than 2,000 ac of forest habitat should receive very little manipulation to maintain necessary habitat features.

A second alternative to manage islands for viable populations and biological diversity would be to establish boundaries for management to include a cluster of islands that would support a viable population within a planning unit. In this alternative, combined area of the island cluster would be equal to that needed to maintain a viable population, given that all islands within the cluster would be within the mean dispersal distance for a species. As mentioned above, dispersal distances vary by species, but may range upward to several miles for large animals, such as deer, bear, or moose. With such dispersal, even one individual per generation is believed adequate to maintain genetic heterozygosity at a level needed to maintain a viable population (Samson et al. 1985).

A third alternative would be an umbilicus of islands that would maintain dispersal between mainland and offshore islands. Given the lack of information on distribution and dispersal abilities for most species in southeast Alaska, the likelihood of establishing island links is limited. The theory and concept behind such an approach, however, has been discussed by Diamond and Gilpin (1983).

DISCUSSION AND SUMMARY

Perhaps no other current topic rivals biological diversity in terms of concern and interest among conservationists and scientists (Roberts 1988). Several conservation groups and scientists argue for critical areas or "hot spots" of diversity managed as a system of preserves. Counter to this approach is a recognition that most species neither live in pristine areas nor in pristine areas of a size and composition adequate to maintain viable populations. Rather, most species live on lands often used for timbering, mining and other resource production. Thus, the future for biological diversity—particularly on public lands managed for multiple use—rests with ecologically sound management of this seminatural matrix.

Resource managers recognize that number, size and juxtaposition of habitat patches created by land management influence whether viable population requirements in regulations to the National Forest Management Act (1976) and other legislation are fulfilled, as well as achieving stated wildlife population objectives that likely will be well above any threshold viability level. Less evident in current land management, however, is concern for another conservation goal: biological diversity.

A useful definition of biological diversity is "To maintain in a healthy state both the species and the ecological processes historically native to a natural landscape" (Wilcove and Samson 1987). A number of plausible concepts—diversity (Samson and Knopf 1982), gap theory (Shugart 1984), landscape ecology (Forman and Godron 1986), hierarchy theory (White 1987) and others—may determine patterns in biological diversity. The conventional wisdom of research to test each concept is complicated by interplay of species, communities, and ecosystems that may be subtle yet important (Urban et al. 1987). Recent examples in conservation do suggest that guidelines, whether or not accepted by one's peers, aid in reaching a consensus on controversial issues (Soule and Simberloff 1986). In this vein, the working group as a consensus offers the following recommendations on a controversial issue—the conservation of rain forest in southeast Alaska—and suggests this approach has application to other ecosystems.

1. Consider the seminatural matrix as a focus in management and identify key biological units, particularly old-growth communities precisely defined in ecological terms.
2. Provide for the distribution of key biological units by physiographic province or geographically defined region at a scale where patterns of natural disturbance are considered.
3. Conduct timber harvest so that (1) some remaining old-growth patches persist in perpetuity. (2) Be "sloppy" in harvesting timber. Leave green trees, brushy areas, and other areas of natural disturbance in place at harvest to provide habitat heterogeneity within otherwise often monotypic second growth stands. (3) Maintain large, continuous blocks of old-growth forest that provide the variety and distribution of habitats needed by wildlife.
4. Consider the uniqueness of the ecosystem and species native to that natural landscape. Southeast Alaska has several unique ecosystems and provides unique challenges to resource managers. For example, how do we maintain viable populations on offshore islands? Many interesting questions persist. Should land managers try to produce all resources—wildlife, timber, etc.—on small land areas everywhere or should larger areas be used depending on management emphasis? For example, perhaps some watersheds should be managed for timber while others remain intact for wildlife. Unique ecosystems may require unique solutions, and consensus by technical working groups provides managers with sound background for their decisions.

It is important to note that change in resource management practices is not new. A decade ago, fishery managers removed debris from stream channels. Today, maintenance of logs and other woody debris in streams and rivers is critical in management. Total fire control has shifted to allow some natural fires to burn and prescribed burning is used regularly in forest, grassland, and wetland management. Habitat management for predators (and even reintroductions) is a significant shift from broad-scale eradication of 2 or more decades ago.

In summary, the Red Queen in Lewis Carroll's *Through the Looking Glass* tells Alice, "Now, here, you see, it takes all the running that you can do to keep in the same place." We, as professional resource managers, must do more than staying in place. As Leopold (1933) wrote in *Game Management*, "All factions, whatever their own differences, should unite to make available the known facts," and "bring all theories susceptible to local trial to the test of actual experience." We hope as a diverse working group that ideas and concepts presented in this paper continue to be given serious consideration and test in the Tongass Land Management Plan revision, thus conserving one of North America's great natural resources.

Literature Cited

Alaback, P. 1988. Endless battles, verdant survivors. Nat. Hist. 97:45–48.

Bourgeron, P. S. 1988. Advantages and limitations: protection of ecosystems. Conserv. Biol. 2:218–220.

Cairns, J. 1986. The myth of the most sensitive species. BioScience 36:670–673.

Carroll, L. 1965. Through the looking glass and what Alice found there. Random House, NY. 127 pp.

Diamond, J. M., and M. E. Gilpin. 1983. Biogeographic umbilici and the origin of Philippine avifauna. Oikos 41:307–321.

Fay, F. H. 1985. Preliminary status survey of selected small mammals. Unpub. mimeo on file USDI Fish and Wildlife Service, Juneau, AK. 42pp.

Franklin, J. F., and R. T. T. Forman. 1987. Creating landscape patterns by forest cutting: ecological consequences and principles. Landscape Ecol. 1:5–18.

Forman, R. T. T., and M. Godron. 1986. Landscape ecology. John Wiley and Sons, NY. 476pp.

Harris, L. D. 1984. The fragmented forest. Univ. Chicago Press, Chicago. 211pp.

Leopold, A. 1933. Game management. Charles Scribners Sons, NY. 481 pp.

Lockie, J. D. 1966. Territory in small carnivores. Symp. Zool. Soc. London 18:143–165.

MacArthur, R. H. and E. O. Wilson. 1967. The theory of island biogeography. Princeton Univ. Press, Princeton, NJ. 203pp.

Mealey, S. P., J. F. Lipscomb, and K. N. Johnson. 1982. Solving the habitat dispersion problem in forest planning. Trans. N. Amer. Wildl. and Nat. Resour. Conf. 47:142–153.

Noss, R. F. 1983. A regional approach to maintain diversity. BioScience 33:700–706.

————. 1987. Protecting natural areas in fragmented landscapes. Nat. Areas J. 7:2–13.

————, and L. P. Harris. 1986. Nodes, networks, and MUM's. Environ. Manage. 10:299–309.

Risser, P. G. 1986. Diversity in and among grasslands. Pages 176–180 in E. O. Wilson, ed., Biodiversity. National Academy Press, Washington, DC. 391pp.

Roberts, L. 1988. Hard choices ahead on biodiversity. Science 241:1759–1761.

Rousch, G. J. 1985. The heritage concept entering the second decade. Nature Conserv. News 35:3–11.

Samson, F. B. 1980. Island biogeography and the conservation of nongame birds. Trans. N. Amer. Wildl. and Nat. Resour. Conf. 45:245–251.

Samson, F. B., and F. L. Knopf. 1982. In search of a diversity ethic for wildlife management. Trans. N. Amer. Wildl. and Nat. Resour. Conf. 47:412–431.

Samson, F. B., and F. Perez-Trejo, H. Salwasser, L. F. Rugerio, and M. L. Scaffer. 1985. On determining and managing minimum population size. Wildl. Soc. Bull. 13:425–433.

Schaffer, M. L., and F. B. Samson. 1985. Population size and extinction: a note on determining critical population size. Amer. Nat. 125:144–152.

Schoen, J. W., D. C. Walmo, and M. D. Kirchoff. 1981. Is a reevaluation of old growth necessary? Trans. N. Amer. Wildl. and Nat. Resour. Conf. 46:531–544.

Schoen, J. W., and M. D. Kirchoff. 1988. Little deer in a big woods. Nat. Hist. 97:52–55.

Shugart, H. H. 1984. A theory of forest dynamics. Springer, NY. 373pp.

Scott, J. M., B. Csuti, J. P. Jacobi, and J. E. Estes. 1987. Species richness. A geographic approach to protecting future biological diversity. BioScience 37:682–702.

Soule, M. E., and D. Simberloff. 1986. What do genetics and ecology tell us about the design of nature reserves. Biol. Conserv. 35:19–40.

Temple, S. A. 1983. Area-dependent changes in the bird distribution and vegetation of southern Wisconsin forests. Ecology 64:1057–1068.

Urban, D. L., and R. V. O'Neil, and H. H. Shugart, Jr. 1987. Landscape ecology. A hierarchial perspective can help scientists understand spatial patterns. BioScience 37:119–127.

U.S.D.A. Forest Service. 1986. Draft supplement to the environmental impact statement for the Regional Guide. U.S.D.A. For. Serv., Portland, OR. 322pp.

————. 1988. Status of the Tongass National Forest. U.S.D.A. For. Serv., Juneau, AK. 76pp.

White, P. S. 1987. Natural disturbance, patch dynamics, and landscape patterns in natural areas. Nat. Areas J. 7:14–22.

Wilcove, D. S. 1985. Nest predation in forest tracts and the decline of migratory songbirds. Ecology 66:1211–1214.

_____. 1988. National Forests. Policies for the future. Protecting biological diversity. The Wilderness Society, Washington, DC. 50pp.

_____, and F. B. Samson. 1987. Innovative wildlife management: listening to Leopold. Trans. N. Amer. Wildl. and Nat. Resour. Conf. 52:327–332.

© ED HOAG '90

6

American Marten: A Case for Landscape-level Management

JOHN A. BISSONETTE, RICHARD J. FREDRICKSON, AND
BRIAN J. TUCKER

W E DISCOVER a basic reality in our quest to protect various segments of wildlife habitat against the losses caused by a single resource user. There is a need to develop new attitudes and perspectives in order to gain a genuine improvement in the way a resource should be utilized.

The old-growth forests of Newfoundland provide its inhabitants with a source of revenue derived from logging. In the process, habitat for the American marten is being destroyed. The American marten has specific habitat requirements, and old-growth forests provide avenues to food supplies in the canopy and on the ground. These forests offer cover from predators and a means to minimize heat loss during severe winters. The marten is one indicator species of habitat conditions within these forests. When marten numbers are low, the forest habitat may be in poor condition.

The dilemma centers on the need to maintain a troubled forest industry and also to protect a shrinking habitat for a threatened species. The challenge to resolve this dilemma is multifaceted. First, more and better-directed research data are needed. Second, this information must be applied through new methods of system-wide management at a landscape level to be effective. Third and finally, both economic planning and ecological thinking must be used to create a new means for integrating old-growth harvesting and habitat protection.

Dr. Bissonette and his colleagues devised a series of research objectives to address this challenge. These objectives focused on: (1) marten utilization of residual stands isolated from continuous old-growth forests and the identification of characteristics that would allow for predicting their use; (2) prey availability and densities within logged old-growth stands; and (3) the effects of logging on the spatial dynamics of martens.

The ultimate goal of these studies is to determine how to manage old-growth forest operations so as to optimize two goals: (1) an economic harvest, and (2) the continued maintenance of marten habitat.

It is apparent that the latter goal is becoming increasingly difficult to achieve since its success is directly dependent upon the maintenance of old-growth forests. It is also apparent that the emphasis on the old-growth forests' economic value as a commodity is being maintained, in part, by federal subsidies. A kind of false economy is in the making regarding these landscapes. Cooperative approaches must integrate private industries' economic interests with governmental concerns for benefits associated with conservation of habitat vital to a threatened species. A shift in thinking about the value of old-growth forests as a viable commodity resource is inevitable. The research conducted to determine American marten habitat perhaps will enable the powers in charge to realize the tentative nature of continuing this enterprise.

INTRODUCTION

This paper reports on the cooperative efforts of research scientists, the Newfoundland government, and provincial resource managers to provide for the continued existence of threatened (COSEWIC-Canadian list) American marten (*Martes americana atrata*) in commercially desirable old-growth spruce-fir forests. Achievement of conservation goals, while simultaneously providing for economic interests, is not necessarily an incompatible endeavor, although previous efforts at reconciling the two often have been less than successful. It is our impression that economic development of natural resources often occurs without adequate planning for resource values. The planning that does occur is usually mandated by federal, state, or provincial policies and regulations. This need not be the model we follow.

Perhaps one of the reasons why solutions to complex resource problems seem so difficult to achieve is that attempts to solve them have been focused at an inappropriate scale. Here we argue that landscape-level management is appropriate and necessary for a growing number of resource conflicts. Using marten in Newfoundland as an example, we show that management strategies can be devised that simultaneously promote marten survival, while maintaining commercial logging interests and a viable timber enterprise that supports the economy of western Newfoundland. We first describe marten habitat needs and how they conflict with local economic interests. We then explore the benefits achieved from a landscape approach to the problem.

American marten once were found throughout the province in forested areas, although perhaps never in great numbers (Bergerud 1969, Snyder 1985). Bangs (1897) reported that "as early as 1870 marten were still common in various parts of the island, but from the increasing . . . value of the fur is annually becoming scarcer." Indeed, at this time across northern North America, marten habitat was steadily reduced and changed by settlement and logging (Strickland and Douglas 1983). For

117

many decades following the record harvests of the late 1840s and mid-1850s (Strickland and Douglas 1983), trapping figures reflected a general decline in marten populations (Novak et al. 1987). By 1930, many areas had been severely depleted, and trapping seasons were closed. For example, in New York, marten were common throughout the state in the 1800s but by the turn of the century were restricted to the Adirondacks (Grant 1903). In Newfoundland, marten were sufficiently rare by 1934 to necessitate closing the trapping season. It has never been reopened. Remnant populations in eastern Newfoundland were probably extinct by 1969 (Bergerud 1969). Marten populations across North America began to rebuild in the mid-1970s (Novak et al. 1987), but Newfoundland populations have remained low. Today, the only remnant viable population is concentrated in old-growth timber found in the Little Grand Lake area of western Newfoundland (Snyder and Hancock 1982, Mayo 1984). A marten study area was created in 1973 by the Newfoundland and Labrador Wildlife Division as a refuge where no trapping or snaring has been allowed. Marten were introduced into the headwaters area of the La Poile River (Mayo 1976a), the Main River area (Mayo 1976b), Siviers Island (Porter 1976), and Terra Nova National Park (Bateman 1985) in attempts to establish other populations. It appears that the introduction to Siviers Island and the La Poile River has been unsuccessful, although marten have been reported from the Main River area since 1976 (J. Hancock 1987. pers. comm.).

STUDY AREA

Habitat Preference

Marten prefer dense, mature coniferous forest or mixed forest with high overstory density (Marshall 1951, de Vos 1952, Francis and Stephenson 1972, Koehler et al. 1975, Clark and Campbell 1976, Koehler and Hornocker 1977, Soutiere 1979, Pulliainen 1981, Hargis 1982, Douglass et al. 1983, Raine 1983, Spencer et al. 1983, Bateman 1986, Snyder and Bissonette 1987). The reasons for this preference are not clearly understood, but overhead cover from predation, prey abundance and availability, and thermoregulatory needs during winter (Buskirk et al. 1987) appear to be involved. Large clearcuts have been shown to be detrimental to marten populations in North America (Campbell 1979, Major 1979, Soutiere 1979, Thompson 1982, Snyder 1984), and Europe (Gravok 1972). Even regenerating clearcuts are used infrequently or avoided. In Wyoming, clearcuts less than 1 year old were not used by marten (Clark and Campbell 1977) and regenerating clearcuts in Maine were used only in the late summer when berries were present (Steventon and Major 1982). Commercially clear-cut areas in Maine supported densities two-thirds less than similar uncut areas (Soutiere 1979). In areas of extensive clearcutting in Newfoundland, Snyder and Bissonette (1987) and Bissonette et al. (1988) found marten concentrated their activities in undisturbed forest.

The Current Problem

During the last 30 years, timber harvesting in western Newfoundland has extended from the Corner Brook area near the paper mill to much more remote stands near the Grand Lake–Little Grand Lake area, the last remaining marten habitat on the island. Corner Brook Pulp & Paper Ltd. currently has access to exemptions within and leases surrounding the marten study area and is cutting that timber. Approximately 4,000 of 133,560 ha of forest is cut annually within a larger marten distribution area. This rate will probably increase within the next few years (Snyder 1985). In 1983, forest accounted for about 87% of the 140-km² area studied by Snyder; 49% of the forest has been cut, 18% was in residual stands ranging from 1 to 270 ha in size, and 35% was in uncut forest. Clearcutting has continued in this area since 1983. Roughly 5.3 km² of the 561 km² of forested area on the Environmental Assessment Area has been cut. Although marten are completely protected within the marten study area, their habitat is not. It is unknown how many years are needed after clearcutting for a site to regenerate adequately for marten, but data show that greater than 23 years are needed in Newfoundland (Snyder 1984).

History of the Corner Brook Mill

The Corner Brook Mill began producing paper in 1925 under a tripartite ownership involving the governments of Newfoundland and Great Britain and the Newfoundland Power and Paper Company, a holding company of Armstrong-Whitworth, the builder of the plant (Horwood 1986). In 1926, the mill was sold to the International Power and Paper Corporation of New York. Eric Bowater, later to become chairman of the Bowater

Companies, became a director of the Newfoundland Power and Paper Company in 1923, as plans were being made to build the Corner Brook mill. Bowater Paper Mills Ltd. was organized in 1923 in Great Britain. At the same time, Bowater organized a sales organization to sell Corner Brook paper as its sole agent throughout the world. In 1938, Bowater brought International Paper's interest in the mill and the name of the International Power and Paper Company of Newfoundland was changed to Bowaters Newfoundland Paper Mills Ltd. Bowaters operated the plant until 1984 when it was sold to Kruger Incorporated of Montreal (Horwood 1986).

Just prior to this, in 1980, an Environmental Assessment Act was passed in Newfoundland. The Minister of Environment for the province has responsibility for administering the Act and can require any activity to be registered, even if not specifically outlined on Schedule 1 of the Act, if he feels there is a potential environmental impact. Whether or not an assessment is required depends upon government review and public screening of the registration (J. Hancock, pers. comm.). In late 1981, the Wilderness and Ecological Reserves Advisory Council requested that forestry projects in the Little Grand Lake area of Newfoundland be registered under the Act because of the importance of the area to marten. In January 1982, the Minister of Environment requested that Bowaters register all logging and associated road building activities in the Little Grand Lake area. Bowaters subsequently submitted a registration document. Based on a review of the document by representatives of the provincial and federal government and the public, the Minister directed Bowaters to prepare an environmental impact statement of their forest harvesting operations in the Little Grand Lakes area. The Minister also established the boundaries of the Environmental Assessment Area for the company. In late 1982, Terms of Reference for the environmental impact statement were approved by the Minister of Environment following a review by the Environmental Assessment Committee and the public. The terms required specific component studies for marten and archeology. Bowaters was in the process of selling the Corner Brook plant, and little was done about the assessment until the company was sold to Kruger. Negotiations between Kruger (Corner Brook Pulp & Paper Ltd.) and the provincial government resulted in a Cabinet level exemption, allowing Kruger to cut 100,000 cords of wood in the Environmental Assessment Area. One condition of the exemption order was that annual cutting plans for the exempted area had to be submitted to the Wildlife Division for approval; another required Kruger to undertake a study in cooperation with the

Wildlife Division to determine the impacts of logging on marten. We completed that work in 1987 (Bissonette et al. 1988). Independent of these activities, we conducted an earlier study of marten habitat selection supported by the Newfoundland and Labrador Wildlife Division in 1982–84 (Snyder 1984, Snyder and Bissonette 1987).

USE OF RESIDUAL FOREST STANDS BY MARTEN

Newfoundland forests are characterized by occasionally steep topography and numerous hydric sites, including bogs, streams, rivers, and other wet areas. Logging operations characteristically leave residual patches of forest varying in size from less than 1 ha to ⩾40 ha. The primary method of tree harvest has been clearcutting of contiguous, large tracks of balsam fir (*Abies balsamea*) and black spruce (*Picea mariana*). Old-growth stands are most profitably harvested. Some of the best stands of old growth left on the island are found in the Grand Lake–Little Grand Lake area. The question we asked was whether marten used residual stands isolated from continuous old-growth forest, and if so, were there identifiable characteristics that would allow us to predict their use.

Field work for our first study (Snyder and Bissonette 1987) was conducted from June 1983 to March 1984. Residual stands were grouped into five size classes and clearcuttings into three categories based on the height of balsam fir regeneration. Live trapping was conducted from June through December in 43 residual stands and 35 clearcuttings. Vegetational characteristics were taken at each trap site. Chi-square and stepwise regression analyses were conducted to determine which variables accounted for the difference between successful and unsuccessful trap sites. Snow tracking was conducted from January through March. Habitat selection was determined by comparing distance of marten trails observed in each habitat with an expected distance based on area.

Captures in Residuals vs. Clearcuttings

Fifty-one (89.5%) marten captures were in residual stands and 6 (10.5%) were in clearcuttings. Mean captures per 100 trap nights were 2.19 in all residual stands and 0.48 in clearcuttings; total trap nights equaled 3,593.

Residual stands 25 to 34.5 ha in size had 4.62 captures per 100 trap nights (Snyder and Bissonette 1987). Marten capture rates differed with residual stands size ($X^2 = 13.36$, df = 4, $P = 0.010$). Only 5 (10%) of 51 captures were in residual stands ≤ 15 ha. Thirteen captures would be expected if marten use was independent of residual stand size (Snyder and Bissonette 1987). Tree height, percentage overhead cover, presence of slash, and distance to nearest habitat edge contributed most to the difference between residual and clearcut trap sites. Tree dbh was the only variable contributing significantly to trap success.

Tracking

Marten tracks were followed for 29 km (Snyder and Bissonette 1987). Seventy-four percent of marten trails were located in forested habitat that comprised 46% of the area. Clearcuttings represented 41% of the area, but only 25% of marten travel was recorded there (Snyder and Bissonette 1987). Clearly, marten prefer residual forest over clearcut and regenerating areas, and will use patches of residual forest if greater than 15 ha. The clear implication is that interior, or core area is required. Our recommendations suggested a change in logging operations from large-scale clearcuttings to much smaller-scale patch cuts. Indeed, our second study, conducted from November 1985 to August 1987, was designed to assess the impact of an altered harvesting regime. Corner Brook Pulp & Paper Ltd. agreed to intersperse clearcuttings into a mosaic of cut and uncut areas. Woods operations began with road building in June 1986; clearcut harvesting began 4 August and ended 7 November, 1986. No harvesting was done in 1987 except for a right-of-way cutting in August to facilitate road building. During the course of the study, 7,860 cords of pulpwood were harvested from 259 ha by clearcutting; 4.3 km of capital road and 2.6 km of forest roads were constructed. We studied the impact of the new forest harvest methods on small mammals, the principal prey base of marten. We also documented home-range spatial dynamics of marten pre- and post-harvesting to determine if significant changes in spatial use were evident.

Prey Base

In Newfoundland, the depauperate small mammal fauna provides a limited prey base for marten. Only 2 of the 7 small mammal species are native: meadow voles (*Microtus pennsylvanicus*) and arctic hares (*Lepus arcticus*). Snowshoe hares (*Lepus americanus*), masked shrews (*Sorex cin-*

erous), eastern chipmunks (*Tamias striatus*), and red squirrels (*Tamiasciurus hudsonicus*) were introduced in 1864, 1958, 1862, and 1963, respectively (Peterson 1966, Northcott 1974, Northcott et al. 1974, Payne 1976). Deer mice (*Peromyscus maniculatus*), first documented by a single specimen found in 1968 (Gould and Pruitt 1969), appear to have become established in small, isolated pockets in western Newfoundland (Tucker et al. 1988). Only 4 of the 7 species (meadow voles, masked shrews, red squirrels, and snowshoe hares) are found in any abundance. Only meadow voles and masked shrews are abundant in old-growth forests of the area. Chipmunks are found in moderate numbers locally in the Barachois Pond Provincial Park area and appear to be spreading throughout the western part of the province. Arctic hares inhabit upland barrens.

Shrews and voles were trapped in control vs. experimental areas in old-growth as well as clearcut areas 3 weeks, 1 year, 13 years and 23 years post-logging, using a trapping-web design with 240 trap sites, 3 traps per site (Anderson et al. 1983). Trapping was conducted before, during, and after logging. The first trapping series occurred in spring 1986 with three webs each placed in a control, as well an experimental area that was cut later. The second series took place in fall of 1986. Three webs were placed in mature timber (control) and in a 3-week-old clearcut. The third and final series occurred in spring 1987. Three webs each were placed in mature timber (control) and in three experimental areas representing 1-year, 13-year, and 23-year cutovers (Bissonette et al. 1988). Small mammal densities were calculated and differences compared with t-tests at $P = 0.05$.

Shrew densities were significantly higher in older cutovers 13 and 23 years of age, but no differences in densities were found between the control old-growth and 3-week and 1-year logging areas. Densities in the control equaled 26.6 animals per ha versus 26.8 for the experimental area prior to logging. At 3 weeks post-logging, mean densities were 15.9 versus 13.4 per ha (control), and 15.7 versus 10.3 per ha (control) 1-year post-logging. Shrew densities increased to 75.1 versus 15.7 per ha (control) in 13-year cutovers, and 51.3 versus 15.7 animals per ha in 23-year cutovers.

Vole densities were higher 3-weeks post-logging, but declined to almost zero in 1-, 3-, 13-, and 23-year-old clearcuts the following spring. It was apparent that some factor other than habitat was involved in the decline because densities in the control areas also declined precipitously. As a result, we were unable to determine how structure in older clearcuts was related to vole densities.

Measures of abundance were obtained for the remaining prey species.

None was found in great abundance. Our data indicated that voles and shrews were found in 91% and 15% of marten scats, respectively (Tucker 1988).

Marten Dynamics

We also studied the effects of logging on marten spatial dynamics. We were interested to learn how logging in established home ranges affected use of that range. Marten were captured, telemetered, and relocated 1 to 3 times daily from June 1986–August 1987. Relocation data for ten marten on the control and experimental areas were grouped into three treatments of home-range use before, during, and after logging, and spatial dynamics analyzed. Three indices were used to test for significant autocorrelation within a data set: Psi (Swihart and Slade 1985a), Gamma (Swihart and Slade 1986), and t^2/r^2 (Swihart and Slade 1985b). A nonvariance supported estimate was made of population size.

On the control area, two females and three males established home ranges. Both females occupied their home ranges until late 1986 and appeared unaffected by logging operations. All males were present throughout the logging. One expanded his range into the experimental area upon the death of a conspecific. None appeared affected by logging. In the experimental area, home range information was obtained for two females, two male kits and one adult male. One female changed her activity pattern shortly before logging, the other shortly after harvest operations. Their movements showed strong avoidance of clearcut areas. Throughout the study, adult marten were found in clearcuts in only 7 of 324 locations (2.2%). The two male kits expanded their range throughout the summer and fall of 1986 before dispersing in the fall. Some home-range expansion by young marten, independent of harvesting operations or other causes, is expected. Contact was lost with the remaining male before harvest.

We estimated resident marten population size based on mean home-range sizes for males and females and the area of available habitat within the Environmental Assessment Area. Within-sex ranges do not overlap appreciably (Bissonette et al. 1988). According to forest inventory charts, there was approximately 561 km^2 of marten habitat within the assessment area. We defined marten habitat as stands of mature softwood and mixed wood. Using harmonic mean estimates, we calculated mean female home-range sizes to be 6.64 km^2 ($n = 4$), and mean male ranges to be 9.19

km² (*n* = 3). Marten home ranges were well established and represented values as close to the true state of nature as is possible with this sample size. Male and female populations were estimated separately because home ranges overlapped between, but not appreciably within, sexes. We calculated that the Environmental Assessment Area can support approximately 150 resident marten; the effective breeding population may be much smaller, depending on sex and age class ratios (Bissonette et al. 1988).

Our results clearly indicate that marten require old-growth habitat and that populations will decline as old-growth timber is cut. In Newfoundland, populations are not dense. Absolute numbers are apparently quite low. An undisturbed interior, or core area, of old-growth must be maintained if marten are to survive. However, economic pressures for cutting, as well as the presence of insect defoliators, make core-area management a dynamic concept and practice.

INTERVENING VARIABLES: POLITICS AND THE ECONOMY

Forest operations in Corner Brook date to 1926 and were predicated on a quick return on capital invested. The result was a rapid consumption of nearby wood and an expansion into more remote regions, including the Grand Lake–Little Grand Lake area. A planned cropping to promote sustained yield never materialized (Horwood 1986). Indeed, the economics of old-growth are predicated upon liquidation. Clawson (1976) has projected that a $12 billion inventory in standing old-growth timber means a potential annual cost, at modest interest rates, of $600 million (Harris 1984). Given the demand for wood pulp, substantial economic pressure to cut old-growth spruce and fir in Newfoundland will continue.

Corner Brook Pulp & Paper Ltd. remains the cornerstone of the economy in western Newfoundland. Any action taken to protect marten or their habitat is mediated by its impact on the economy of the region. Apart from small manufacturing and the service trades, the city of Corner Brook is wholly dependent on the paper industry (Horwood 1986). The industry has tended to be cyclic, not necessarily responding in synchrony with the economy but rather to surpluses in milled paper goods. The years 1926, 1937, 1958, the early 1970s, and 1982–1983 were bad economic times for the Corner Brook mill (Horwood 1986). In 1982–1983, the mill closed its largest of several machines and 800 people were laid off. The province already was suffering high unemployment. High mortgage rates for homeowners and high bank rates for businessmen exacerbated the situation. Building and home construction was at a low ebb.

When the mill was sold to Krueger in 1984, the federal and provincial governments provided generous subsidies, and the unions representing forest products workers made substantial concessions. The province provided $64 million: $30 million in loan guarantees to the banks if Krueger failed, the remaining $34 million in direct and indirect subsidies. Thirty-three million was matched by the federal government under the Federal-Provincial Pulp and Paper Modernization Agreement (Horwood 1986). The union accepted restricted benefits in exchange for job security. Additionally, a hotly contested amendment to the Labor Standards Act (Bill 37) was passed, retroactively shortening the period required for notice of temporary layoffs to 1 week. The company saved $6.67 million in employee layoff payments before purchasing the mill (Horwood 1986).

Any actions resulting in the removal of significant amounts of old-growth conifer timber in marten habitat will significantly reduce marten numbers. The literature is replete with documentation. Knowledge of marten habitat requirements is not the limiting factor. However, the reality in western Newfoundland is that cutting will continue. Old-growth will be removed. The challenge is to orchestrate woods operations in such a manner as to allow an economic harvest while yet providing for the continued existence of threatened marten and the other species dependent on large expanses of old-growth. However, the scale at which current management practices are conducted does not promote the maximum likelihood of success. The areas involved are insufficient in size, given the annual volume of wood required to run the mill. Estimates of the area of remaining old-growth habitat in western Newfoundland are not available, but according to our calculations (Bissonette et al. 1988), approximately 561 km^2 of old-growth remain within the Environmental Assessment Area, where the greatest marten densities occur. Corner Brook Pulp & Paper Ltd. has cut approximately 4,000 ha of forest annually in the decades prior to 1986 in an area of approximately 134,000 ha, stretching south from the Main River area to Barachois Brook. Snyder (1985) suggests that this rate is likely to increase in the future. Actual harvest will no doubt be mediated by the world supply of milled paper. If the 11-year Corner Brook cycle holds, demand for wood pulp will be high into the mid-1990s. The long rotation times needed to attain the characteristic structure of old-growth spruce and fir in Newfoundland, even with silvicultural practices, suggests that a good deal of planning is necessary to effectively manage simultaneously for wood production and for maintenance and enhancement of marten habitat. Once removed, old-growth may require 80 to 100 years to regenerate. The logistics and expense of recovering large tracts of timber killed by spruce-budworm (*Choristoneura fumiferana*), as well as other primary and secondary organisms (Raske and Sutton 1986), is part of the equation. Forest management units in western Newfoundland rated very high on a vulnerability index to spruce budworm defoliation (Raske 1986). Dead timber is commercially usable for only 2 to 3 years after defoliation. In western Newfoundland, tree mortality from spruce budworm increased from 137,950 cords in 1981 to 881,225 cords by 1983. Total stand volume containing these trees was 2,847,840 cords over 95,700 ha (Raske and Sutton 1986). It is clear that management on a much larger scale is required.

LANDSCAPE MANAGEMENT

The principles of landscape ecology and management provide the necessary tools. Landscapes have been defined as heterogeneous land areas composed of a cluster of interacting ecosystems, or landscape elements, occurring repeatedly across the land area (Forman and Godron 1986). Landscape-level management presupposes knowledge of ecological properties, the physical and biological relationships that govern the different spatial units in the landscape. Patches, corridors, and the background matrix of vegetation are the units of interest and measure.

Forman and Godron (1986) have suggested that landscape formation is a result of three processes: (1) specific geomorphological processes; (2) colonization patterns of organisms; and (3) the disturbance of regimes of individual ecosystems over a shorter time. Landscape ecology, therefore, is similar to systems ecology, addressing structure, functions, and their change over time. However, the landscape approach focuses horizontally on relationships between spatial units, rather than vertically within a spatial unit (Forman and Godron 1986).

The landscape in western Newfoundland is composed not only of old-growth and regenerating areas, but also of bogs, barrens, high-elevation heath, ponds, and lakes, as well as riparian communities. Marten move across these landscape elements, utilizing vegetation patches and corridors, as well as interior areas of old-growth when seeking cover and food. The core of the problem, then, is how to orchestrate logging with marten habitat requirements across the landscape.

The first step is to represent the entire landscape in a computer interactive format, i.e., a system (GIS) allowing easy and interactive measurement of the distribution of landscape elements, or tesserae (Forman and Godron 1986), including clearcuts, regenerating forest, and old growth. Bogs, barrens, and other landscape features also are included and assessed for spatial characteristics. Forest type maps can be used to provide the initial map, or better still, satellite imagery, using the most time-appropriate scene.

Quantitative measures of the spatial properties of patches, corridors, and matrices in a GIS-enhanced landscape provide the basis for assessing habitat quality and the tools for region-wide management. For example, once the type of patch has been determined (disturbance, regenerating, remnant, environmental resource; Forman and Godron 1986), size (area) measurements are simple and informative. Snyder and Bissonette (1987) found that marten are sensitive to patch size of residual forest stands. Diversity of other species also correlates closely with patch size (Whitcomb 1977, Robbins 1980, Ambuel and Temple 1983). Patch shape, measured as the ratio of the perimeter of the patch to the circumference of a circle of identical area, has important biological implications for marten. Temple (1986) has shown graphically how core area varies with patch shape. Patches that tend toward minimum perimeter length (i.e., isodiametric shapes) maximize interior area and minimize edge. Our data (Snyder and Bissonette 1987) and that of others demonstrate that marten populations are denser in undisturbed forests with large core areas. Residual forest patches ⩾ 15 ha, and with patch shape values tending toward unity, would appear to be desirable landscape elements for marten if logging must ultimately result in greater habitat heterogeneity. Patch isolation and degree of connectedness also may be important considerations (Fahrig and Merriam 1985). Corridor curvilinearity (measured as a ratio of corridor length to straight line length), and connectivity (a measure of the number of breaks of specified width) in an otherwise continuous habitat strip are properties that appear to have biological importance for species that require continuous cover but exist in patchy environments. Finally, the relative area of the matrix, the most connected and extensive landscape element (Forman and Godron 1986), is perhaps the most critical requirement for core-sensitive species. One can envision a landscape where the perturbation is so extreme that the patches created by the disturbance become the matrix by virtue of their prevalence and connectedness. For core-area species, the most preferred habitat must be the landscape matrix.

Landscape-level management of marten in Newfoundland can be reduced to a series of related questions based on the following guidelines: (1) old-growth should be the matrix element in the landscape and clearcut patches should be kept within specified sizes; (2) if residual patches result from cutting they should be no smaller than a specified minimum size with isodiametric shapes; and (3) corridor access routes, preferably along riparian corridors, should be maintained between patches. Additional guidelines regarding spatial arrangement of landscape elements should be incorporated into the management plan.

Some of the questions we can ask are: (1) What is the maximum size of clearcut allowed? (2) What total amount of wood can be cut per year and still maintain enough old-growth with sufficient interior area to support healthy populations of martens? (3) What are the optimum locations for timber removal to maximize spatial arrangements of habitat elements? (4) What is the rotation time from clearcut to characteristic old-growth structure? We suggest that the answers to most of these questions are relatively easy to extract, using existing data. The more broad-scale questions can be addressed with interactive GIS-based maps.

Finally, transition matrices provide a quantitative way to model change in a landscape. Transition matrices are nothing more than a series of lines of algorithms representing replacement rates for various parameters and arrayed as a table (Forman and Godron 1986). For instance, rotation time, amount of old-growth removed each year, amount remaining, and proportion of regenerating forest in specific age classes might be integrated into one transition matrix. A second might include the dynamics of insect-killed softwood i.e., its mean rate of appearance within the landscape, expected length of time it has wood value for the mill, and expected removal rate. Two or more matrices can be displayed in an array of several dimensions called a tensor (Johnson and Sharpe 1976, Franklin 1979, Kassell and Potter 1980). The outcome is a predictive model that can be used to orchestrate wood removal on a landscape scale. A helpful property of transition matrices is that, regardless of the initial proportions of the various habitat elements that one begins with, if replacement rates remain constant over time, viz., several years of constant wood removal, the proportions of the various habitat elements converge toward a stable equilibrium. It is possible to iterate the methodology and let the biology of marten determine what those optimum proportions are *a priori*.

Landscapes, especially those with significant disturbance regimes, are dynamically changing habitat mosaics. Attempts to perpetuate old-growth in localized areas present nearly insurmountable problems. Insect-caused needle and cone diseases, fungal tree pathogens, wind-throw, and other life-history-related mortality factors all contribute to tree loss, causing structural changes in the landscape. Likewise, the forces of vegetational succession are constantly at work, and significant areas revert to regenerating forest. No old-growth stand is exempt. As a result, management aimed at preserving a specific patch of old-growth shows little promise of success. Landscape-level management is needed. Management across a large landscape eliminates the need to protect any one stand in perpetuity. Rather, the landscape elements are managed to maintain a shifting mosaic with predetermined proportions of each habitat element. To the extent we understand the habitat needs of core-sensitive species such as marten, we can mimic the spatial arrangements of critical elements to provide for those needs under a high-disturbance regime of cutting.

SUMMARY

Environmental assessment legislation, a cyclic resource-based economy, and a perennially scarce species provide the need for a new approach to management of old-growth forest in eastern Canada. We report the results of our work and make the case that management of timber resources in Newfoundland is inextricably intertwined with marten management and must be planned and conducted on a landscape scale. GIS methodology is a necessary component. Management practices designed with nature must provide the underlying philosophy. Long-range landscape planning is required if marten survival and economically feasible timber harvests are to be achieved sympatrically over the long term. Marten management is integrated vertically throughout Newfoundland society, culture, and politics—from the logger in the field and his concerns to put bread on the table, to the paper mill at Corner Brook and its emphasis on annual profits, to the provincial government's interest in providing for a thriving resource-based economy in western Newfoundland, to the civil servants in the Divisions of Wildlife and Environment, who are charged with conserving the environment and its biota. Laws and regulations that protect wildlife and certain habitat values are in place. We have provided recommendations on how diverse and sometimes contradictory interests can be reconciled under existing legislation by using concepts and precepts from the emerging field of landscape ecology.

Acknowledgments

We thank M. E. Bissonette, J. A. Chapman, G. S. Drew, and T. C. Edwards for helpful criticism in preparing this paper.

Literature Cited

Ambuel, B., and S. A. Temple. 1983. Area dependent changes in the bird communities and vegetation of southern Wisconsin forests. Ecology 64:1057–1068.

Anderson, D. R., K. P. Burnham, G. C. White, and D. L. Otis. 1983. Density estimations of small mammal populations using a trapping web and distance sampling methods. Ecology 64:674–680.

Bangs, O. 1897. Preliminary description of the Newfoundland marten. Amer. Nat. 31:161–162.

Bateman, M. C. 1985. Termination report on the Atlantic region reintroduction program. Rep. prepared for Parks Canada by Can. Wildl. Serv., Sackville, N.B., Canada. 10pp.

————. 1986. Winter habitat use, food habits, and home range size of marten, *Martes americana*, in western Newfoundland. Can. Field-Nat. 100:58–62.

Bergerud, A. T. 1969. Status of pine marten in Newfoundland. Can. Field-Nat. 83:128–131.

Bissonette, J. A., R. J. Fredrickson, and B. J. Tucker. 1988. The effects of forest harvesting on marten and small mammals in western Newfoundland. Rep. prepared for the Nfld. & Labr. Wildl. Div., and Corner Brook Pulp & Paper Co., Ltd. Utah Cooperative Fish and Wildl. Res. Unit, Utah State Univ., Logan. 109pp.

Buskirk, S. W., H. J. Harlow, and S. C. Forrest. 1987. Management of subalpine forests: building on 50 years of research. Pages 150–153 *in* C. A. Troendle, M. R. Kaufmann, R. H. Hamre, and R. P. Winokur, tech. coord. U.S.D.A. For. Serv. Gen. Tech. Rep. RM–149.

Campbell, T. M. 1979. Short-term effects of timber harvests on pine marten. MS Thesis. Colorado State Univ., Fort Collins. 71pp.

Clark, T. W., and T. M. Campbell. 1976. Population organization and regulatory mechanisms of pine marten in Grand Teton National Park, Wyoming. First Conf. Res. in Nat. Parks, New Orleans. 9pp.

Clark, T. W., and T. M. Campbell. 1977. Short-term effects of timber harvest on pine marten behavior and ecology. Unpubl. Term. Rep., Idaho State Univ., Pocatello. Unnumbered.

Clawson, M. 1976. The national forests. Science 191:762–67.

de Vos, A. 1952. The ecology and management of fisher and marten in Ontario. Tech. Bull. Ontario Dep. Lands, For., Wildl. Ser. 1. 90pp.

Douglass, R. J., L. G. Fisher, and M. Mair. 1983. Habitat selection and food habits of marten, *Martes americana*, in the Northwest Territories. Can. Field-Nat. 97:71–74.

Fahrig, L., and G. Merriam. 1985. Habitat patch connectivity and population survival. Ecology 66:1762–1768.

Forman, R. T. T., and M. Godron. 1986. Landscape ecology. John Wiley & Sons. NY. 619pp.

Francis, G. R., and A. B. Stephenson. 1972. Marten ranges and habits in Algonquin Provincial Park, Ontario. Minis. Nat. Resour. Rep. (Wildl.) 91. Toronto, Canada. 53pp.

Franklin, J. F. 1979. Ecosystem studies in the Hoh River drainage Olympic National Park. Pages 1–8 *in* E. E. Starkey, J. F. Franklin, and J. W. Mathews, eds., Ecological research in national parks of the Pacific Northwest. For. Res. Lab. Pub., Oregon State Univ., Corvallis.

Gould, W. P., and W. O. Pruitt Jr. 1969. First Newfoundland record of *Peromyscus*. Can. J. Zool. 47:469.

Grakov, N. M. 1972. Effects of concentrated woods fellings on the abundance of the pine marten (*Martes martes* L.). Byull. Mosk. O-va Ispyt. Prin. Otd. Biol. 77:14–23. *in* Soutiere, E. C. 1978. Effects of forest management on the marten in Maine. Ph.D. Thesis. Univ. Maine, Orono. 62pp.

Grant, M. 1903. Notes on Adirondack mammals with special reference to furbearers. Pages 319–334 *in* New York Game and Fish Commission Rep. 1902–1903. Albany.

Hargis, C. D. 1982. Winter habitat utilization and food habits of pine marten in Yosemite National Park. Tech. Rep. No. 6, Coop. Natl. Park. Resour. Stud. Unit, Univ. Calif., Davis. 59pp.

Harris, L. D. 1984. The fragmented forest: island biogeography theory and the preservation of biotic diversity. Univ. Chicago Press, Chicago. 211pp.

Horwood, H. 1986. Corner Brook: a social history of a paper town. Newfoundland History Series No. 3. Breakwater Books Ltd., St. John's, Nfld., Canada. 182pp.

Johnson, W. C., and D. M. Sharpe. 1976. An analysis of forest dynamics in the northern Georgia piedmont. For. Sci. 22:307–322.

Kassell, S. R., and M. W. Potter. 1980. A quantitative succession model for nine Montana forest communities. Environ. Manage. 4:227–240.

Koehler, G. M., and M. G. Hornocker. 1977. Fire effects on marten habitat in the Selway-Bitterroot Wilderness. J. Wildl. Manage. 41:500–505.

Koehler, G. M., W. R. Moore, and A. R. Taylor. 1975. Preserving the pine marten: management guidelines for western forests. West. Midl. 2:31–36.

Major, J. T. 1979. Marten use of habitat in a commercially clearcut forest during summer. M.S. Thesis. Univ. Maine, Orono. 48pp.

Marshall, W. H. 1951. Pine marten as a forest product. J. For. 49:899–905.

Mayo, L. 1984. Pine marten distribution study in Newfoundland, 1983. Intern. Prog. Rep. No. 3081, Newfoundland and Labrador Wildl. Div., Pasadena, Nfld., Canada. 11pp.

————. 1976a. Introduction of pine marten to the head of the La Poile River. Intern. Prog. Rep. No. 75PM-2. Newfoundland and Labrador Wildl. Div., St. John's, Nfld., Canada. 4pp.

————. 1976b. Transfer of pine marten from Grand Lake to Main River. Intern. Prog. Rep. No. 76PM-1. Newfoundland and Labrador Wildl. Div., Pasadena, Nfld., Canada. 11pp.

Northcott, T. H. 1974. The land mammals of insular Newfoundland. Newfoundland Dept. Tourism. St. John's Nfld., Canada. 90pp.

————, E. Mercer, and E. Menchenton. 1974. The eastern chipmunk, *Tamias striatus*, in insular Newfoundland. Can. Field-Nat. 88:86.

Novak, M., M. E. Obbard, J. G. Jones, R. Newman, A. Booth, A. J. Satterthwaite, and G. Linscombe. 1987. Furbearer harvests in North America, 1600–1984. Ontario Minist. Nat. Resour., Ont. Trappers Assoc., North Bay, Ontario, Canada. 270pp.

Payne, N. F. 1976. Red squirrel introductions to Newfoundland. Can. Field-Nat. 90:60–64.

Peterson, R. L. 1966. The mammals of eastern Canada. Oxford Univ. Press, London. 465pp.

Porter, B. 1976. Pine marten introduction to Siviers Island. Intern. Prog. Rep. No. 76PM-3. Newfoundland and Labrador Wildl. Div., St. John's, Nfld., Canada. 6pp.

Pulliainen, E. 1981. Winter habitat selection, home range, and movements of the pine marten (*Martes martes*) in a Finnish Lapland forest. Pages 1068–1089 *in* J. A. Chapman and D. Pursley, eds., Proc. Worldwide Furbearer Conf., Frostburg, MD.

Raske, A. G. 1986. Vulnerability rating of the forests of Newfoundland to spruce budworm damage. Nfld. For. Ctr. Inform. Rep. N-X-239. Can. For. Serv., St. John's, Nfld., Canada. 16pp.

————, and W. J. Sutton. 1986. Decline and mortality of black spruce caused by spruce budworm defoliation and secondary organisms. Nfld. For. Ctr. Inform. Rep. N-X-236. Can. For. Serv., St. John's, Nfld., Canada. 29pp.

Raine, R. M. 1983. Winter habitat use and responses to snow cover of fisher and marten in southeastern Manitoba. Can. J. Zool. 61:25–34.

Robbins, C. S. 1980. Effect of forest fragmentation on bird populations. Pages 198–212 *in* DeGraaf, R. M., and K. E. Evans, compilers. Management of north central and northeastern forests for nongame birds. U.S.D.A. For. Serv. Gen. Tech. Rep. NC-51.

Snyder, J. E. 1984. Marten use of clearcuts and residual forest stands in western Newfoundland. M.S. Thesis. Univ. Maine, Orono. 31pp.

Snyder, J. E. 1985. The status of pine marten (*Martes americana*) in Newfoundland. Unpub. Rep. prepared for COSEWIC and the Newfoundland and Labrador Wildl. Div., St John's, Nfld., Canada. 35pp.

Snyder, J. E., and J. A. Bissonette. 1987. Marten use of clear-cuttings and residual forest stands in western Newfoundland. Can. J. Zool. 65:169–174.

Snyder, J. E., and J. Hancock. 1982. Pine marten investigations. Nfld. & Labr. Wildl. Div., St. John's, Nfld., Canada. Int. Rep. 38pp.

Soutiere, E. C. 1979. Effects of timber harvesting on marten in Maine. J. Wildl. Manage. 43:850–860.

Spencer, W. D., R. H. Barrett, and W. J. Zielinski. 1983. Marten habitat preferences in the northern Sierra Nevada. J. Wildl. Manage. 47:1181–1186.

Steventon, J. D., and J. T. Major. 1982. Marten use of habitat in a commercially clear-cut forest. J. Wildl. Manage. 46:175–182.

Strickland, M. A., and C. W. Douglas. 1983. The marten. Ontario Minist. Nat. Resour., Toronto, Canada. 14pp.

Swihart, R. K., and N. A. Slade. 1985a. Influence of sampling interval on estimates of home-range size. J. Wildl. Manage. 49:1019–1025.

Swihart, R. K., and N. A. Slade. 1985b. Testing for independence of observations in animal movements. Ecology 66:1176–1184.

Swihart, R. K., and N. A. Slade. 1986. The importance of statistical power when testing for independence in animal movements. Ecology 67:255–258.

Temple, S. A. 1986. Predicting impacts of habitat fragmentation on forest birds: a comparison of two models. Pages 301–304 in J. Verner, M. L. Morrison, and C. J. Ralph. eds., Wildlife 2000: modeling habitat relationships of terrestrial vertebrates. Univ. Wisconsin Press, Madison.

Thompson, I. D. 1982. Effects of timber harvesting of boreal forest on marten and small mammals. Prog. Rep. No. 1. Can. Wildl. Serv., Ottawa. Ont., Canada. 22pp.

Tucker, B. J. 1988. The effects of forest harvesting on small mammals in western Newfoundland and its significance to marten. M. S. Thesis. Utah St. Univ., Logan. 49pp.

————, A. Bissonette, and J. Brazil. 1988. Deer mouse, *Peromyscus maniculatus*, in insular Newfoundland. Can. Field-Nat. 102:722–723.

Whitcomb, R. F. 1977. Island biogeography and "habitat islands" of eastern forests. Amer. Birds 31:3–5.

© ED HOAG '90

7

Planning for Basin-level
Cumulative Effects in the
Appalachian Coal Field

WILLIAM C. McCOMB, KEVIN McGARIGAL,
JAMES D. FRASER, AND WAYNE H. DAVIS

L ANDSCAPE-PLANNING ISSUES can be viewed simultaneously as a problem to be solved and a challenge to plan for success. Such is the case when considering the problems created by intensive coal mining and forest logging in the coal fields in Kentucky. Deforestation took place in the early 1900s. Since then, more than 1-million acres have been surface mined for coal. Dr. McComb and others estimate that if all remaining coal resources in Kentucky were removed more than 1.3 million acres of commercial forestland would be affected.

McComb suggests that the transition from an economy dependent upon coal to one dependent on timber and other resources is inevitable. His proposal is to start now in order to plan for the eventual conversion to a renewable resource-based economy. This economy will be based upon more realistic expectations of sustainable yields.

McComb and his colleagues project the obvious impacts likely to be associated with intensive forest logging: (1) the reduction of acreage covered with mature forests, and (2) fragmentation of mature-forest stands. Their recommendations include a three-part plan to minimize the cumulative effects of resource exploitation that have accumulated over the last century. They recommend: (1) reforestation of reclaimed surface-mining areas; (2) careful planning of location and size of timber harvests so as to minimize reduction of core areas of adjacent mature stands; and (3) maintenance of mature-forest stands in the managed landscape by extending rotations to 150 to 200 years.

Because resource exploitation in this region of the country has resulted in a ravaged and depleted landscape, it might be difficult for those who have knowledge of this devastation to take these recommendations seriously. Yet these recommendations have merit and are worthy of our serious attention. These recommendations also demonstrate a basic principle of any resource-planning action: Resource planning is either reactionary or revolutionary in its character. Earlier reactionary decisions simply responded to the availability of a resource with little regard to future consequences. An entire economy and social order were constructed around these resource decisions.

Now, a new and revolutionary ideas are being proposed that will build a new social order based on a more sustainable resource-based economy. To give these plans consistency, they are based upon ecological and landscape-planning theory that respects the realities of sustainable resource development and recovery.

McComb and his coauthors are simply recognizing the problems and impacts realistically. Their work is an attempt to mesh a people's desire to live off the land while still preserving it.

138

INTRODUCTION

Land-use patterns in the Appalachian coal field are likely to change in the next few decades. The cumulative effects of these land uses on wildlife could result in dramatic changes in wildlife community structure. In this paper, we define cumulative effects as the disproportionate increase or decrease in a wildlife population with linear changes in areal habitat availability. Of particular concern is the potential decrease in abundance of species associated with mature forest stands (sawtimber stands at or beyond sawtimber rotation age) due to the depletion and fragmentation of mature forests by surface mining and timber harvesting. Results among studies examining fragmentation influences on wildlife are quite consistent: large blocks of contiguous mature forest support disproportionately more species and more individuals of some species than small forested blocks (Anderson and Robbins 1981). Neotropical migrant warblers inhabiting mature forests may be particularly susceptible to the effects of forest fragmentation. If cumulative effects are important, then land-use changes in the near future could have large-scale impacts on some wildlife species.

We propose that a basin (or watershed) is the logical planning unit for assessing cumulative effects of land uses on wildlife. Basins provide habitat for both aquatic and terrestrial species, and these habitats are linked by hydrologic and colluvial processes. Further, transportation systems in the region usually follow valleys or ridges, so these systems make logical boundaries for management units.

In this paper, we will describe current and expected land-use patterns in the Appalachian coal field, discuss the potential cumulative effects of anticipated land-use changes, and suggest research needs and approaches to minimize those impacts on wildlife. While our discussion will focus on the Appalachian coal field, we believe that many of our inferences are relevant to other portions of the eastern hardwood forest.

THE APPALACHIAN COAL FIELD

We define the Appalachian coal field as that portion of the mixed meso-phytic forest that lies within the Cumberland Plateau and Cumberland Mountains from West Virginia to north Georgia and north Alabama. Coal mining currently is concentrated in the northern portion of the region. The area encompasses approximately 10 million ha. Approx-imately 94% of the land is privately owned, but parts or all of five national forests and many state forests and state parks lie within the region (Austin 1965).

More than 230 species of terrestrial vertebrates occur in the region. The mixed mesophytic forest has the richest floral, breeding bird, mammal, and amphibian communities of any upland eastern U.S. forest type (Hinkle et al. 1989). A high percentage (>60%) of the breeding bird community is composed of neotropical migrants.

PAST, PRESENT, AND FUTURE LAND USES

Recent U.S.D.A. Forest Service inventories of timber resources in the portion of the central Appalachians underlain by coal (e.g., Craver 1985) lead us to believe that three dominant types of disturbance are likely to occur in the region over the next 20–30 years: (1) surface mining on ridge tops and side-slopes; (2) moderate to high density, single-family housing development along valley bottoms; and (3) harvesting of mature hard-wood forest on midslopes and in coves. This combination of forest distur-bance has not occurred previously in the Appalachians despite past resource exploitation.

Human Settlement

Prior to European settlement, human populations were low (1 per square mile), and some slash and burn agriculture was practiced (Hinkle et al. 1989). Human habitation has always been concentrated in valleys. Consequently, most larger valleys and bottomlands have been disturbed by housing or farming (Barber 1984), and this land use will likely continue despite depressed economies.

Forest Resources

Because disturbance regimes in old-growth eastern hardwood forests are of small scale but high frequency (predominantly windthrow), many animal species likely evolved to inhabit landscapes dominated by mature or old-growth forest. Prior to the 1900s, the forest was comprised predominantly of mature oaks (*Quercus* spp.), hickories (*Carya* spp.), yellow-poplar (*Liriodendron tulipifera*), American chestnut (*Castanea dentata*), American beech (*Fagus grandifolia*), eastern hemlock (*Tsuga canadensis*), and pines (*Pinus* spp.) (Braun 1950). Timber cutting began in the early 1800s, but forests still comprised 50% to 60% of the land area through the nineteenth century (Hinkle et al. 1989). After 1870, however, timber and coal resource exploitation increased rapidly. In 1889, 61% of West Virginia was forested; by 1910, uncut forests represented only 10% of the land area. By 1930, most old-growth forests were cut. Currently, second-growth forests comprise about 80% of the land area (Hinkle et al. 1989). Both sawtimber volume and acreage of sawtimber stands are increasing in the region (Fig. 7.1.), and most stands are 40–80 years old (Fig. 7.2.). However, because of poor wood product markets, poor transportation systems, and local economies dominated by coal and/or tourism, the harvest of growing stock is decreasing (Kingsley and Powell 1977, Craver 1985, Brown 1986*a,b*). Consequently, sawtimber acreage and growing stock are expected to continue increasing for the next 20 years (Kingsley and Powell 1977, Bones 1978). Most stands will be at or beyond rotation age by that time. Although forest products industries currently comprise 15% to 20% of some local economies (Schallau et al. 1985, Maki et al. 1987), the contribution to most local economies is low (<10% of the economic base), and Schallau et al. (1986*a,b*) believe that it will remain low until the next century.

Coal Mining

Since the extensive deforestation of the region in the early 1900s, the coal industry has dominated the regional economy. More than 405,000 ha have been surface mined in the region (Kingsley and Powell 1977, Bones 1978).

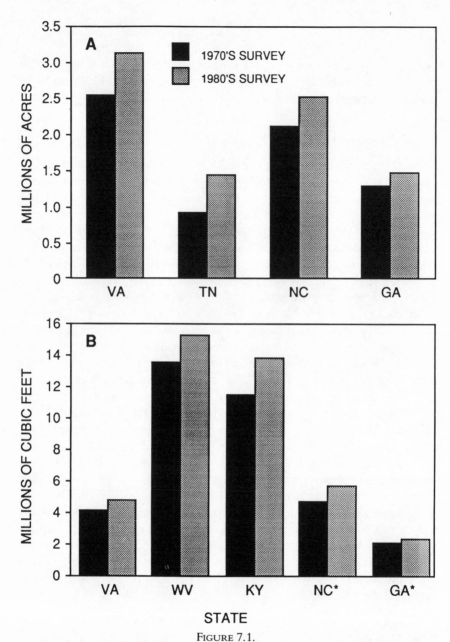

FIGURE 7.1.
Changes in sawtimber acreage (A) and volume (B) from 1970s to 1980s for selected states in the Appalachian coal region, based on U.S.D.A. Forest Service inventory data (Bones 1978, Kingsley and Powell 1977, Brown 1986a,b, Craver 1985, Cost 1974, Knight 1972, Sheffield 1977a,b, Tansey 1983, V. A. Rudis, pers. comm.).

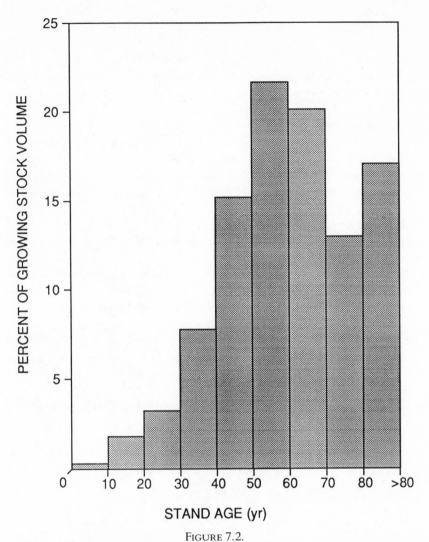

FIGURE 7.2.
Distribution of growing stock volume (N=4,828,597 million cubic feet) by stand age class for the mountain regions of North Carolina and Virginia, based on U.S.D.A. Forest Service 1985 inventory data (Craver 1985, Brown 1986a,b).

Surface mine acreage in Kentucky increased exponentially from the mid-1950s to the mid-1970s (Kingsley and Powell 1977). If all extractable coal is removed from Kentucky's eastern and western coal fields, then roughly 527,000 ha of commercial forest land in Kentucky will be affected (Kingsley and Powell 1977).

Coal is mined using both deep mining and surface-mining techniques. Surface mining results in deforestation followed by mountain-top removal or bench mining along contours. The goal of reclamation is to establish vegetative cover on the site within 2 years in order to minimize erosion. A common practice is to reclaim to grasses (particularly *Festuca arundinacea*) and legumes and then graze the lands. The result is an area dominated by exotic herbaceous plants and patches of black locust (*Robinia pseudoacacia*).

Coal is a non-renewable resource. As coal reserves are exhausted, local economies will either collapse or turn to other resources, such as wood products and tourism. The timber resource will be of an age and size appropriate for harvest within the next 20 years. Transportation systems already will be established along valleys and ridge tops in surface-mined areas, thereby minimizing one constraint on forest management (Kingsley and Powell 1977). We believe that the eventual decline in the coal resource, coupled with the growing potential revenue in hardwood products, will cause these forests to be harvested with increasing intensity within the next few decades.

CUMULATIVE EFFECTS OF LAND USES ON WILDLIFE

There are at least two mechanisms that could lead to decreased habitat quality for species associated with mature forests as a result of surface mining and timber harvesting: reduction of mature-forest acreage and fragmentation of mature-forest stands.

Mature-Forest Wildlife

Continued surface mining and intensive timber management in the future could significantly reduce the abundance and distribution of mature forests and reduce the abundance of wildlife populations dependent on mature forests. Recent studies have documented the importance of mature forests to some wildlife species. McGarigal and Fraser (1984) investigated the influence of forest-stand age on great horned owl (*Bubo virginianus*) and barred owl (*Strix varia*) distributions in southwestern Virginia. They documented higher response rates to recorded owl vocalizations in old stands (>80 years) than young stands (<80 years).

Pileated woodpeckers (*Dryocopus pileatus*) (Conner et al. 1975) and several other relatively common species may also be closely associated with mature forests (Odum 1950). Mature, mixed mesophytic forests also provide habitat for species such as red-cockaded woodpeckers (*Picoides borealis*) (Mengel 1965) and common ravens (*Corax corax*) (Fowler et al. 1985). Undoubtedly, some of these species were affected adversely by logging of the area in the early 1900s.

Wildlife associations with old-growth forests have recently been documented in the Pacific Northwest (Meslow et al. 1981). Little old-growth remains in eastern hardwood forests, and few studies have been conducted in mature eastern hardwood forests (Carey 1983, Rosenberg et al. 1988). We suggest the need for additional studies in the Appalachians to further identify species dependent upon mature forests and their components and to determine the nature of those dependencies. Additionally, we must determine whether the truncated age distribution of forest stands caused by logging and surface mining is detrimental to these species.

Forest Fragmentation

Intensive timber management and surface mining in the Appalachian coal field not only will reduce the amount and distribution of mature forest, but also will produce a fragmented landscape. Mature-forest stands will become smaller and more isolated as they become imbedded in a mosaic of young-forest stands (created by timber harvesting) and grassland corridors (created by surface mining). The resulting decrease in core areas of mature-forest stands and increase in edge could have dramatic effects on the avian community (Gates and Gysel 1978, Temple 1986).

The creation of early seral stage and edge habitats will undoubtedly benefit some wildlife species (McComb 1985, Yahner and Howell 1975), and at moderate levels will probably increase local and regional vertebrate diversity. White-tailed deer (*Odocoileus virginianus*) are abundant in portions of the coal region where clearcutting has increased the proportion of early seral-stage habitat. Similarly, species such as white-footed mice

(*Peromyscus leucopus*), short-tailed shrews (*Blarina brevicauda*), and white-eyed vireos (*Vireo griseus*) would probably benefit by creation of early seral stage patches in mature forest (McComb 1985). In addition, extensive grasslands formed following surface-mine reclamation provide habitat for some species, such as grasshopper sparrows (*Ammodramus savannarum*), bobolinks (*Pooecetes gramineus*), and prairie voles (*Microtus ochrogaster*), that had not previously occurred in the mixed mesophytic forest (Claus et al. 1988, Barbour and Davis 1974).

On the other hand, some wildlife species associated with mature forests will be affected adversely. Clearcutting adjacent to mature forest and thinning of mature forest led to decreases in abundances of ovenbirds (*Seiurus aurocapillus*) (Fig. 7.3.). Similarly, Webb et al. (1977) and Robbins (1984) reported ovenbirds to be sensitive to forest management and forest fragmentation, respectively. Yahner and Howell (1975) reported that ovenbirds preferred mature forest over surface mine edges, and Allaire (1978) reported a decrease of ovenbirds from 15.2 per 40 ha to 0.4 per 40 ha following creation of a surface mine edge adjacent to mature forest. Kentucky warblers (*Oporornis formosus*), worm-eating warblers (*Helmitheros vermivorous*), black-and-white warblers (*Mniotilta varia*), and black-throated green warblers (*Dendroica virens*) were similarly affected. Densities of breeding birds decreased 18% one year after the mining (Allaire 1978).

Recent observations suggest that brood parasitism by brown-headed cowbirds (*Molothrus ater*) (Brittingham and Temple 1983) and nest predation by edge-dwelling predators (Yahner and Scott 1988, Wilcove 1985), such as common crows (*Corvus brachyrhynchos*), striped skunks (*Mephitis mephitis*), opossums (*Didelphis virginianus*), black racers (*Coluber constrictor*), and rat snakes (*Elaphe obsoleta*), may be responsible for population declines in mature-forest species. In Kentucky, brown-headed cowbirds occurred (1–3 birds/10 ha) in thinned stands and along clearcut edges, but not in mature 60-year-old forests (McComb, unpubl. data). Allaire (1978) reported a slight increase in brown-headed cowbird abundance from 4.1 to 5.3 per 40 ha following creation of a surface mine edge along a mature forest. Claus et al. (1988) reported flocks of 500 to 600 brown-headed cowbirds on surface mines reclaimed to grassland. Neotropical migrant warblers inhabiting adjacent mature forests could suffer particularly high levels of brood parasitism (Gates and Gysel 1978, Brittingham and Temple 1983). We suggest that reforestation of surface mines through reclamation should be considered as an alternative to grassland reclamation on surface mines adjacent to mature forests.

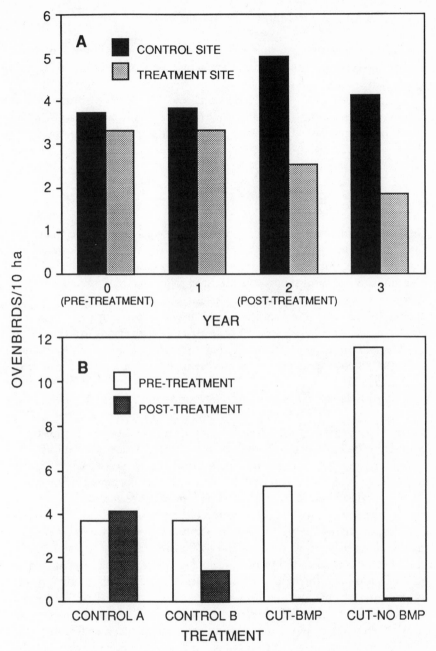

FIGURE 7.3.
Ovenbird response to timber-stand improvement (TSI) cutting (A), and Best Management Practices (BMP) clearcutting (B; control A was a remote site located in the forest interior, control B was adjacent to a clearcut), Robinson Forest, Breathitt County, Kentucky, 1983–1986.

These studies suggest that surface mining, coupled with intensive timber harvesting, could have detrimental effects on some vertebrate species and could affect the structure of vertebrate communities. We believe that additional research is needed to determine the optimal distribution of seral, mature, and old-growth forest stands on the landscape to minimize cumulative effects and to assess the comparative merits of alternative reclamation strategies.

MINIMIZING CUMULATIVE EFFECTS

Given the dominance of mature forest in the region, resource managers have the opportunity to proactively plan for the needs of mature-forest species. While much remains to be learned about optimal areal and distribution targets of seral stages, we offer the following recommendations to land managers planning future land uses.

First, encourage reforestation of reclaimed surface mines, especially those adjacent to mature forest. Reforestation with native tree species is feasible, and tree growth can be rapid on some sites (Plass 1975). Reduction of grassland cover and grazing would reduce habitat quality for brown-headed cowbirds, and regenerating forests may act as buffers between existing grasslands and mature forests (Askins and Philbrick 1987).

Second, carefully plan the location and size of timber harvests so as to minimize reduction in core area of adjacent mature stands. A minimum fragmentation approach of concentrating harvest in one basin while leaving an adjacent basin in mature forest until the harvested basin has regrown may be a viable alternative (Franklin and Forman 1987), particularly when structural attributes of old forests are retained in harvest units (large snags, large logs, scattered old trees).

Third, maintain mature forest stands in managed landscapes by extending rotations beyond 80 years to 150–200 years, and identify stands that should be left unharvested to produce old-growth. Although mature forests will dominate the landscape in the near future, very few old-growth forests remain. Linkages between mature forest stands may be important (MacClintock et al. 1977), but will be difficult to attain in this region. Harris (1984) suggested using riparian areas to link mature forests in the Pacific Northwest. This option is generally not available in the Appalachian coal field because of high human density in the valleys. These linkages would only be feasible in this region following land-use legislation that would limit human activities in riparian areas, such as along streams designated for protection under the Wild and Scenic Rivers Act.

Fourth and finally, base land-use decisions on a regional plan that adequately considers optimal amounts and distributions of seral stages from a wildlife-habitat perspective. Success in developing such a plan will require a large-scale study on the optimal spatial arrangements of different habitat components and the optimal scale for planning and management. However, even with the necessary information in hand, implementation of a regional management plan will be complicated by the patchwork of jurisdictions and surface and mineral ownerships in the region. It may be possible to overcome these difficulties by implementing a coordinating council and a program of landowner incentives. The former could be patterned after the waterfowl flyway councils and would consist of federal and state officials and other interest groups. Thus, the council would ensure that local interests were not overlooked in creating goals and policies for the region as a whole. Landowner incentives could be patterned after the Conservation Reserve Program and would be aimed at filling habitat gaps that could not be filled by managing public lands alone.

Acknowledgments

We thank D. F. Stauffer, S. A. Bonney, and C. L. Chambers for reviewing an early draft of this manuscript. This is paper no. 2511 of the Oregon Forest Research Laboratory.

Literature Cited

Allaire, P. N. 1978. Effects on avian populations adjacent to an active strip-mine site. Pages 232–240 *in* D. E. Samuel, J. R. Stauffer, C. H. Hocutt, and W. T. Moon, eds., Proceedings, Surface mining and fish/wildlife needs in the eastern United States. FWS/OBS-78/81. U.S.D.A. Fish and Wildl. Serv., Washington, DC.

Anderson, S. H., and C. S. Robbins. 1981. Habitat size and bird community management. Trans. N. Amer. Wildl. and Nat. Resour. Conf. 46:511–519.

Askins, R. A., and M. J. Philbrick. 1987. Effect of changes in regional forest abundance on the decline and recovery of a forest bird community. Wilson Bull. 99:7–21.

Austin, M. E. 1965. Land resource regions and major land resource areas of the United States, exclusive of Alaska and Hawaii. Handb. No. 296. U.S.D.A., Washington, DC. 82pp.

Barber, H. L. 1984. Eastern mixed forest. Pages 345–354 *in* L. K. Halls, ed., White-tailed deer ecology and management. The Stackpole Company, Harrisburg, PA and The Wildlife Management Institute, Washington, DC.

Barbour, R. W., and W. H. Davis. 1974. Mammals of Kentucky. Univ. Press of Kentucky, Lexington. 322pp.

Bones, J. T. 1978. The forest resources of West Virginia. Resour. Bull. NE-56. U.S.D.A. For. Serv., Upper Darby, PA. 105pp.

Braun, E. L. 1950. Deciduous forests of eastern North America. Blakiston Press, Philadelphia. 596pp.

Brittingham, M. C., and S. A. Temple. 1983. Have cowbirds caused forest songbirds to decline? BioScience 33:31–35.

Brown, M. J. 1986a. Forest statistics for the northern mountains of Virginia, 1986. Resour. Bull. SE-85. U.S.D.A. For. Serv., Asheville, NC. 56pp.

_____. 1986b. Forest statistics for the southern mountains of Virginia, 1986. Resour. Bull. SE-86. U.S.D.A. For. Serv., Asheville, NC. 55pp.

Carey, A. B. 1983. Cavities in trees in hardwood forests. Pages 167–184 in J. W. Davis, G. A. Goodwin, and R. A. Ockenfels, tech. coords., Snag habitat management: proceedings of the symposium. Gen. Tech. Rep. RM-99. U.S.D.A. For. Serv., Fort Collins, Co.

Claus, D. B., W. H. Davis, and W. C. McComb. 1988. Bird use of eastern Kentucky surface mines. Ky. Warbler 64:39–43.

Conner, R. N., R. G. Hooper, H. S. Crawford, and H. S. Mosby. 1975. Woodpecker nesting habitat in cut and uncut woodlands in Virginia. J. Wildl. Manage. 39:144–150.

Cost, N. D. 1974. Forest statistics for the mountain region of North Carolina, 1974. Resour. Bull. SE-31. U.S.D.A. For. Serv., Asheville, NC. 33pp.

Craver, G. C. 1985. Forest statistics for the mountains of North Carolina, 1984. Resour. Bull. SE-77. U.S.D.A. For. Serv., Asheville, NC. 50pp.

Fowler, D. K., J. R. MacGregor, S. A. Evans, and L. E. Schaaf. 1985. The common raven returns to Kentucky. Amer. Birds 39:852–853.

Franklin, J. F., and R. T. T. Forman. 1987. Creating landscape patterns by forest cutting: ecological consequences and principles. Landscape Ecol. 1:5–18.

Gates, J. E., and L. W. Gysel. 1978. Avian nest dispersion and fledging success in field-forest ecotones. Ecology 59:871–883.

Harris, L. D. 1984. The fragmented forest: island biogeography theory and the preservation of biotic diversity. Univ. Chicago Press, Chicago. 211pp.

Hinkle, C. R., W. C. McComb, and S. J. Marcus, Jr. 1989. Mixed mesophytic forest. Chapter 14. in W. H. Martin, ed., The biotic communities of the southeastern U.S. Wiley Publ. Co., NY. (in press).

Kingsley, N. P., and D. S. Powell. 1977. The forest resources of Kentucky. Resour. Bull. NE-54. U.S.D.A. For. Serv., Upper Darby, PA. 97pp.

Knight, H. A. 1972. Forest statistics for north Georgia. Resour. Bull. SE-25. U.S.D.A. For. Serv., Asheville, NC. 34pp.

MacClintock, L., R. F. Whitcomb, and B. L. Whitcomb. 1977. Evidence for the value of corridors and minimization of isolation in preservation of biotic diversity. Amer. Birds 31:6–12.

Maki, W. R., C. H. Schallau, B. B. Foster, and C. H. Redmond. 1987. Tennessee's forest products industry: performance and contribution to the state's economy, 1970 to 1980. Res. Pap. PNW-RP-386. U.S.D.A. For. Serv., Portland, OR. 22pp.

McComb, W. C. 1985. Habitat associations of birds and mammals in an Appalachian forest. Proc. Southeast Assoc. Fish and Wildl. Agencies 39:420–429.

McGarigal, K., and J. D. Fraser. 1984. The effect of forest stand age on owl distribution in southwestern Virginia. J. Wildl. Manage. 48:1393–1398.

Mengel, R. M. 1965. The birds of Kentucky. Amer. Ornithologists' Union Monogr. No. 3. 581pp.

Meslow, E. C., C. Maser, and J. Verner. 1981. Old-growth forest as wildlife habitat. Trans. N. Amer. Wildl. and Nat. Resour. Confer. 46:329–335.

Odum, E. P. 1950. Bird populations of the highlands (North Carolina) plateau in relation to plant succession and avian invasion. Ecology 31:587–605.

Plass, W. T. 1975. An evaluation of trees and shrubs for planting surface-mine spoils. Res. Pap. NE-317. U.S.D.A. For. Serv., Upper Darby, PA. 8pp.

Robbins, C. S. 1984. Management to conserve forest ecosystems. Pages 101–107 *in* W. C. McComb, ed., Proceedings, Workshop on management of nongame species and ecological communities. Dep. of Forestry, Univ. Kentucky, Lexington.

Rosenberg, D. K., J. D. Fraser, and D. F. Stauffer. 1988. Use and characteristics of snags in young and old forest stands in southwest Virginia. For. Sci. 34:224–228.

Schallau, C. H., W. R. Maki, B. B. Foster, and C. H. Redmond. 1985. North Carolina's forest products industry: performance and contribution to the state's economy, 1970 to 1980. Res. Pap. PNW-343. U.S.D.A. For. Serv., Portland, OR. 22pp.

———. 1986*a*. Kentucky's forest products industry: performance and contribution to the state's economy, 1970 to 1980. Res. Pap. PNW-354. U.S.D.A. For. Serv., Portland, OR. 22pp.

———. 1986*b*. Virginia's forest products industry: performance and contribution to the state's economy, 1970 to 1980. Res. Pap. PNW-368, U.S.D.A. For. Serv., Portland, OR. 22pp.

Sheffield, R. M. 1977*a*. Forest statistics for the northern mountain region of Virginia, 1977. Resour. Bull. SE-41. U.S.D.A. For. Serv., Asheville, NC. 33pp.

———. 1977*b*. Forest statistics for the southern mountain region of Virginia, 1977. Resour. Bull. SE-42. U.S.D.A. For. Serv., Asheville, NC. 33pp.

Tansey, J. B. 1983. Forest statistics for north Georgia, 1983. Resour. Bull. SE-68. U.S.D.A. For. Serv., Asheville, NC. 29pp.

Temple, S. A. 1986. Predicting impacts of habitat fragmentation on forest birds: a comparison of two models. Pages 301–304 *in* J. Verner, M. L. Morrison, and C. J. Ralph, eds., Wildlife 2000: modeling habitat relationships of terrestrial vertebrates. Univ. Wisconsin Press, Madison.

Webb, W. L., D. F. Behrend, and B. Saisorn. 1977. Effect of logging on songbird populations in a northern hardwood forest. Wildl. Monogr. No. 55. The Wildlife Society, Washington, DC. 35pp.

Wilcove, D. S. 1985. Nest predation in forest tracts and the decline of migratory songbirds. Ecology 66:1211–1214.

Yahner, R. H., and J. C. Howell. 1975. Habitat use and species composition of breeding avifauna in a deciduous forest altered by strip mining. J. Tenn. Acad. Sci. 50:142–147.

Yahner, R. H., and D. P. Scott. 1988. Effects of forest fragmentation on depredation of artificial nests. J. Wildl. Manage. 52:158–161.

8

Breeding Bird Assemblages in Managed Northern Hardwood Forests in New England

RICHARD M. DEGRAAF

© ED HOAG '90

S OME CONCEPTS of resource development and conservation can never be built into any land-management scheme. Any landscape has only a limited range of land-use practices that are acceptable to its people. This notion bears importantly on the conditions for managing the forest mosaic of New England.

Meeting demands for multiple use of these forests has been the charge of the U.S.D.A. Forest Service for almost three decades. The public wishes to realize sustained yields not only of forest products but also of increased wilderness and recreation opportunities, improved water quality, and improved wildlife-habitat values. The U.S.D.A. Forest Service has been attempting to optimize these values within their management plans for years but with limited success for at least two reasons. First, New England forests have changed from primeval forest to cleared agricultural land to a mosaic of town-farm-forest landscape over a span of 350 years. Human-caused stresses to the forest communities have resulted in large-scale disturbances from which they can never recover. As a result, higher productivity levels, improved water quality, increased recreational opportunities, and greater wildlife habitats have suffered.

Second, virtually all of the remaining forests are not in a true state of climax but rather in a state of anthropogenic climax. The surviving plant communities result from a prolonged type of land use perpetrated by human occupants. It is unlikely that these forests will realize increased productivity and renewed diversity and stability until human use and occupation are decreased or removed.

This is the setting in which Dr. DeGraaf has conducted his work. His primary objective is to develop guidelines for the selection and management of vegetative structure that support bird populations. Operating under the hypothesis that vegetative structure is the most important factor affecting habitat selection, DeGraaf attempts to identify those management activities that impact bird habitats on either a landscape scale (percent forest cover) or a smaller scale (crown volumes).

Birds play a variety of critical roles in the maintenance of forest ecosystems. DeGraaf's extensive work on habitat relationships of breeding birds is of paramount importance to forested landscapes in New England.

INTRODUCTION

New England's woodlands provide a variety of habitats that support a wide range of wildlife communities. Now mostly forested, the New England landscape has changed dramatically during the last 350 years. Once covered by primeval forest, land was cleared for agriculture— slowly until about 1750, then at an increased pace until, between 1820 and 1840, 75% of the arable land in southern and central New England was in farm crops and pasturage. One hundred years later, New England was about 75% forested—the result of an era of farm abandonment that began in 1830 with the opening of the rich farmlands of Ohio via the Erie Canal. The building of railroads, the Civil War, and even the California gold rush each contributed to the exodus of farmers from the stony hills so arduously brought under cultivation.

This chapter describes changes that occur in bird species composition as stands of northern hardwoods develop after clearcutting in northern New England; this is an example of the changes that occur in one habitat type in the landscape mosaic. Breeding bird compositions associated with different habitats within the northern hardwood type are included, as are some habitat-management guidelines to enhance bird species richness in managed stands and over the landscape as a whole.

The White Mountains of northern New Hampshire and western Maine contain a variety of forest cover types; northern hardwood types predominate and compose approximately half the forested landscape. Most mature stands in the White Mountains originated after extensive fires that were fueled by logging debris in the late nineteenth century. A few older stands, including stands of virgin timber, remain. Younger stands have resulted from even-aged management; clearcutting is the most common method of regeneration.

New England northern hardwood sites are characterized by some combination of sugar maple (*Acer saccharum*), American beech (*Fagus grandifolia*), and yellow birch (*Betula alleghaniensis*). Other associates are red maple (*Acer rubrum*), eastern hemlock (*Tsuga canadensis*), white ash (*Fraxinus americana*), northern red oak (*Quercus rubra*), eastern white pine (*Pinus strobus*), balsam fir (*Abies balsamea*), and red spruce (*Picea rubens*). Quaking aspen (*Populus tremuloides*), bigtooth aspen (*P. grandidentata*), paper birch (*Betula papyrifera*), gray birch (*B. populifolia*), and black cherry (*Prunus serotina*) or pin cherry (*P. pennsylvanica*) are pioneer species that usually occur in mixtures after fire or clearcutting. Over time, they are normally replaced by more shade-tolerant northern hardwoods.

In northern New England, the type provides potential habitat for a total of 126 bird species comprised of 115 breeding and 40 wintering species. By comparison, the red maple type provides potential habitat for 114 breeding and 53 wintering species, and the red spruce type provides for 92 breeding and 39 wintering birds (DeGraaf and Rudis 1986).

In northern hardwood forests, as with other hardwood types, bird species composition varies with timber size-class, stand area, the presence or absence of softwoods in the stand, within-stand features such as cavity trees, openings, and wet spots, and the presence of understory and mid-canopy vegetation layers.

Vegetation structure is the most important factor affecting habitat selection by forest birds (Hilden 1965, James 1971, Anderson and Shugart 1974). Further resource partitioning (niche separation) through specific habitat selection allows breeding birds of different species to coexist in temperate forests (Noon and Able 1978). The set of resources partitioned along gradients or dimensions related to forest structure is referred to as the habitat niche. It is this set of habitat features or components that forest birds perceive when selecting breeding habitat. Some habitat features that affect the potential diversity of bird species present can be manipulated by the land manager.

Birds are excellent candidates for management on a landscape scale because they are habitat-specific in terms of vegetation. Even in human-dominated areas, New England breeding bird diversity is largely a function of tree-crown volume at small scales (Goldstein et al. 1986) and of percent of forest cover at the landscape scale (DeGraaf et al. 1976). In forested regions, breeding birds are closely related to forest cover types and timber size-classes (DeGraaf and Chadwick 1987).

Birds and Vegetation Structure

Lack (1933) first proposed the idea that birds select breeding habitats by responding to features that they do not actually require for survival, namely, the vegetation structure. Many studies have attempted to identify the features or patterns sought by different bird species (e.g., Anderson and Shugart 1974, Willson 1974). These generally indicate that many of the factors (e.g., food availability for nestlings) that ultimately determine reproductive success are not evident to a bird at the time of arrival or habitat selection. These *ultimate* factors are implied by *proximate* factors— aspects of the physical environment, particularly aspects of the vegetation structure, that birds recognize when settling on their breeding grounds.

Studies of habitat selection and partitioning have included measurement of many features of stand structure (e.g., canopy height, closure, tree size-class, understory height, and ground cover) in an attempt to identify which stand attributes represent proximate factors that birds select.

The vertical complexity of forest vegetation (i.e., the diversity of vegetation heights and the density of foliage at those heights) is associated with breeding forest-bird diversity (MacArthur and MacArthur 1961). This relationship of bird species diversity to foliage height diversity generally has been observed in many different forest conditions (Karr 1968, Karr and Roth 1971, Willson 1974).

Horizontal diversity or patchiness, such as a distribution of timber size classes and openings, is also important to breeding bird composition. Horizontal habitat heterogeneity has been shown to be better than vertical habitat heterogeneity for predicting bird species numbers (Roth 1976).

The close relationship between habitat structure and bird species composition is useful for assessing the effects of forest management on breeding birds at both stand and landscape scales. Both the vertical diversity or structure of forest stands and the distribution of stands of different size-classes or types can be manipulated in forest management. Both can be altered to manage bird habitat type and availability.

MANAGED STANDS AND BREEDING BIRDS

Breeding bird species composition varies with timber size-class. This is especially evident in pure stands of even-aged northern hardwoods, and dramatically so in young stands. Breeding bird surveys have been conducted in the White Mountains of New Hampshire in regenerating (0 to 5 yrs), seedling-sapling (6 to 15 yrs), poletimber (16 to 79 yrs), sawtimber

(80–125 yrs), large sawtimber (>125 yrs), and uneven-aged stands to evaluate differences in bird species composition. These surveys were conducted within the interiors of stands that were at least 10 ha in size (DeGraaf 1987).

Analysis of the survey data revealed four northern hardwood habitats that have different breeding bird compositions: regenerating, sapling, pole, and sawlog, and mature stands. The most dramatic differences occur in the smallest-size classes, and breeding bird composition is essentially unchanged in stages beyond the pole-timber stage (Fig. 8.1a. and Fig. 8.1b.). Even-aged sawtimber and large sawtimber stands and uneven-aged stands have similar avifaunas. Species richness (number of species) is similar in regeneration/seedling, sapling, and sawtimber stands. Pole-timber stands have the fewest bird species (DeGraaf 1987).

Bird Species Changes after Clearcutting

In winter-logged stands larger than 10 ha, breeding bird composition changes quite rapidly. In the first growing season after complete removal of all live stems, white-throated sparrows, winter wrens, and willow flycatchers are generally abundant (see Appendix 8.1. for scientific names of birds). Winter wrens are associated with heavy slash and willow fly-catchers with brushy habitats. If stubs with old woodpecker holes are left, eastern bluebirds are commonly present; northern flickers are usually present, either nesting in stubs, scattered trees containing deadwood, or in such trees on the edge of the cut.

Bird species composition can be enhanced by leaving stubs and poten-tial live-cavity trees in clearcuts. In fact, the time required for many species to appear in developing even-aged stands can be shortened appre-ciably if clearcuts are not treated to remove unmerchantable stems down to 5 cm dbh (Table 8.1.).

Two years after clearcutting, the number of species doubles: common yellowthroats, chestnut-sided warblers, cedar waxwings, American goldfinches, and mourning warblers invade, along with Swainson's thrushes, rufous-sided towhees, and American redstarts. Northern flickers and white-throated sparrows are still present, but eastern blue-birds and sometimes winter wrens are not.

In the third growing season after clearcutting, bird species numbers again double with about a dozen new species added, mostly in low numbers. During the next 12 years, bird species composition changes

TABLE 8.1.

Breeding Bird Occurrence, Expressed in Number of Years After Clearcutting, in Northern Hardwoods, White Mountains, NH (after DeGraaf 1987)

Bird Species	First Appear	Become Common	Decline
	In "Clean" Clearcuts		
Willow flycatcher	1	2	5–7
Winter wren	1	4	7–10
White-throated sparrow	1	2	[a]
Swainson's thrush	2	4	15
Chestnut-sided warbler	2	4	10
Mourning warbler	2	5	7–10
Common yellowthroat	2	6	10
Cedar waxwing	2	4	7–10
American goldfinch	2	6	7–10
Veery	3	10	20
Black-and-white warbler	3	10	[a]
Rose-breasted grosbeak	3	15	[a]
Canada warbler	5	15	[a]
Ruffed grouse (drumming males)	10	15	20
Red-eyed vireo	10	[a]	[a]
Wood thrush	10	15	[a]
Ovenbird	10	15	[a]
Black-throated blue warbler	15	[a]	[a]
Scarlet tanager	25	[a]	[a]
	If Stubs Left		
Eastern bluebird	1	1	2
Northern flicker	1	1	7–10
Tree swallow	1	1	7–10
	If Stubs, Trees, Left Near Seep, Wet Spot		
Olive-sided flycatcher	1	1	3–4
	In Untreated Clearcut		
Swainson's thrush	1	1	15[a]
Canada warbler	2	4	[a]
Downy woodpecker	5	[a]	[a]
Black-throated blue warbler	5	[a]	[a]

[a] Present until next cutting cycle

| | Even-aged | | | | | | | | | Uneven-aged |
| | Regeneration | | | | | Sapling | | | Pole | Sawlog | Mature | |
	1	2	3	4	5	6	10	15	16-79	80-125	>125	
Eastern Bluebird	•											
Ruby-throated Hummingbird			•	•	•	•						•
Swainson's Thrush		•	•	●	•	•	•				•	•
Chestnut-sided Warbler		•	●	●	●	●	•	•			•	
Mourning Warbler		•	•	●	●	●	•					
Common Yellowthroat	●	●	•	●	●	●	•					
American Redstart		•	•	●	•	●	●	•	•	•	•	•
American Goldfinch	•	•		•	•	●						
Rufous-sided Towhee	•											
White-throated Sparrow	•	•	•	•	•	●	•	•	•	•	•	
Cedar Waxwing		●	•	•	•	•	•					
Winter Wren	•									•	•	
Northern Flicker	•	•	•	•	•	•						
Willow Flycatcher	•	●	●	●	●	•						
Blue Jay						•	•			•	•	•
Black-capped Chickadee						•	●		•	•	•	•
Gray Catbird		•	•	•		•	•					
Veery		•	•	•		●	●	•	•		•	•
Red-eyed Vireo		•	•	•	•	•	•	•	●	●	●	●
Philadelphia Vireo		•	●	●		•						
Black-and-White Warbler		•	●	●		•	•	•		•	●	•
Nashville Warbler		•	•	•		•	•	•				
Rose-breasted Grosbeak		•	●	●		•	●	•	•	•	•	●
Indigo Bunting	•											
Dark-eyed Junco		•									•	

FIGURE 8.1a. AND FIGURE 8.1b.

Typical breeding bird occurrence in even-aged and uneven-aged New England northern hardwood stands (• = present, ● = common, ⬤ = abundant). After DeGraaf (1987).

| | Even-aged | | | | | | | | | Uneven-aged |
| | Regeneration | | | | | Sapling | | | Pole | Sawlog | Mature | |
Species	1	2	3	4	5	6	10	15	16 - 79	80 - 125	> 125	
Song Sparrow					•							
Eastern Kingbird					•							
Canada Warbler					•	•	•	●			•	•
Black-billed Cuckoo						•	•					
Hermit Thrush						•				•	•	•
Ruffed Grouse (drumming male)						●	•		•	•	•	•
Wood Thrush							•	●	●	•	•	●
Ovenbird							•	●	●	●	●	●
Least Flycather								•		●	•	
Black-throated Blue Warbler								•	●	●	•	•
Yellow-rumped Warbler								•	•	•	•	
Black-throated Green Warbler								•	•	•	•	●
Yellow-bellied Sapsucker									•	•	•	●
Scarlet Tanager									●	•	•	•
Eastern Wood-Pewee										•	•	•
White-breasted Nuthatch										•	●	•
American Robin										•	●	•
Solitary Vireo										•	●	•
Blackburnian Warbler										●	•	
Purple Finch										•	•	•
Hairy Woodpecker										•	•	•
Downy Woodpecker									•	•	•	•

significantly, but the number of species usually does not change appreciably. New species replace those whose habitat requirements are no longer satisfied.

Breeding bird composition changes quite rapidly in the first 10 to 15 years after complete clearcutting. Many of the earliest-arriving birds decline in just a few years as habitat conditions change. In the White Mountains, regenerating stands 1 to 5 years old contain about 28 bird species. Of these, five are restricted to that stage. Sapling stands contain about 30 species, and pole-timber stands only about half as many. The bird communities of sawtimber stands, large sawtimber stands, and uneven-aged stands are not different and contain an average of 30 species (DeGraaf 1987). Large sawtimber stands are generally about 125 years old, just beyond commercial rotation age. In northern New England, old-growth conditions probably begin several hundred years after clearcutting (Bormann and Likens 1979). Virgin stands of northern hardwoods do not contain any bird species that are not found in younger stands (Absalom 1988).

Mature northern hardwood stands in New England commonly contain softwoods, usually red spruce, eastern hemlock, or white pine. As a result, they also contain species that are associated with coniferous forests, such as red-breasted nuthatches, golden-crowned kinglets, and northern parulas, among others. In the absence of conifers, these birds normally would not inhabit a hardwood stand.

Even-aged management in New England northern hardwoods provides habitat for more forest bird species than does uneven-aged management. A range of even-aged stands of northern hardwoods—seedling, sapling, pole timber, and sawtimber—contains twice as many bird species as do extensive uneven-aged stands. All species that occur in the uneven-aged stands also occur in one or more size-classes of even-aged stands. No breeding birds are restricted to uneven-aged conditions. However, many species are restricted to even-aged habitats, especially in regeneration and sapling stands (DeGraaf 1987).

There are essentially no differences in foliage profiles in northern hardwood stands more than 30 years old (Aber 1979). This probably accounts for the similarities of breeding bird assemblages in stands beyond pole-timber size. Foliage profiles in stands of widely disparate ages are similar because northern hardwood species reach their maximum height of about 25 m fairly early in the life of the stand. Also, short trees of shade-tolerant species will persist in the understory even after the canopy closes.

Even-aged sawtimber and uneven-aged stands have similar diameter distributions (Leak 1979). Precommercial thinning in even-aged stands and partial cutting in uneven-aged stands produce similar diameter distributions (Leak 1964, 1965, Carey and Healy 1981, Healy 1987). Once an even-aged stand is beyond pole-timber size, its structure (and thus its bird habitat value) is similar to that of an uneven-aged stand. From a bird habitat perspective, only at final harvest of the even-aged stand would the choice of management system be apparent.

Mature northern hardwood stands have breeding bird compositions that are rather constant under a wide range of canopy conditions, from undisturbed to heavily disturbed. Almost all bird species' populations return to levels characteristic of uncut stands 10 years after disturbance. Postlogging responses of breeding birds are short lived in relation to the long intervals between stand treatments in northern hardwoods (Webb et al. 1977).

Even-aged management creates periodic disturbances that maintain early successional trees, such as yellow birch and aspen in the stands. Birds searching for insects in the canopies of northern hardwood stands do not seem to use tree species randomly, even when trees are similar in life-form. All 10 foliage-gleaning bird species in the Hubbard Brook watershed showed preference for yellow birch, and 7 species (least flycatcher, rose-breasted grosbeak, American redstart, Philadelphia vireo, black-capped chickadee, black-throated green warbler, and Blackburnian warbler) foraged most frequently on yellow birch. Beech was rather uniformly avoided, except by scarlet tanagers. Scattered conifers were used extensively by solitary vireos, Blackburnian warblers, and black-throated green warblers. Most species showed an aversion to beech and sugar maple. Food densities (higher on yellow birch) and adaptations to particular foliage structures are considered the major factors in tree species foraging preferences (Holmes and Robinson 1981).

From a wildlife management standpoint, beech should not be selected against, even though it is not used by foliage-gleaning birds. Beech is the major mast tree throughout most of the range of northern hardwoods. Further, it decays rapidly, readily producing a substrate for cavity users.

Aspen stands of several size-classes are important to ruffed grouse at different seasons of the year. Sapling stands provide secure habitat for drumming males, whereas mature aspens provide buds and catkins for winter food, and broods use recent clearcuts for feeding. Aspen need not be dominant or even commercially valuable to be important as the focus of management for grouse (Gullion 1984). Retaining an early successional species like aspen and benefiting associated wildlife in northern hard-woods is probably best achieved through even-aged management.

Differences in breeding bird assemblages across edges between even-aged stands of northern hardwoods do not persist for long after clear-cutting. New England clearcuts generally do not develop grass or herb stages or brushy borders, but rather regenerate rapidly. Consequently, no distinct breeding bird communities are associated with the edges between forest stands of various size-classes as they are with forest-field edges. The abrupt edge between a clearcut and a mature stand is ephemeral and separates distinct breeding bird communities in each habitat for a brief period—about 12 to 15 years (DeGraaf, unpubl.).

Cavity-Tree Management

In New England northern hardwood forests, 24 bird species nest, roost, or forage for invertebrates in standing trees that have decayed wood (Table 8.2.). Extensive decay in live-tree boles is termed rotten cull by timber managers. Cull trees have little market value, so usually one of the first steps in managing timber stands is to eliminate them. The distribution of rotten cull shifts from large butt logs to upper stems and larger branches of the crowns under increasingly intensive timber management (Cooley 1964, Filip 1978). Cull formation is a slow process, and timber manage-ment can reduce the amount of cull to very low levels over a 10- to 20-year period (Trimble 1963, Tubbs 1977).

Unthinned stands of central New England oaks contain about three times as many cavity trees per unit area as thinned stands, but it is possible to conduct thinnings that both improve timber-stand quality and retain cavity trees because most cull trees do not contain cavities (Healy et al. 1989). Many cavity-nesting birds forage over the boles and limbs of sound trees of all sizes, but they must nest in relatively large trees with suitable decay. For woodpeckers, excavation of the nest hole is vital to pair-bonding at the onset of the breeding season. Habitat needs during the breeding/young rearing period is critical to maintaining populations of these and other species over time.

The number of cavity trees to be retained or provided depends upon the habitat requirements of cavity-dwelling species, site capability to produce large cavity trees, current stand composition and tree condition, and rotation and cutting cycle length.

About one large-diameter (at least 45 cm dbh) cavity/den tree per 2 ha is needed for populations of larger-bodied species, such as wood ducks,

TABLE 8.2.
Types of Trees and Cavities Used by Birds in New England Northern Hardwood Forests (adapted from DeGraaf 1984)

Species	Use[a]	DS 20	DH 15–30	DH 31–45	LCD 20–30	LBTL 30–45	LBTL 45
Wood duck	N						2[c]
Common goldeneye	N						2
Hooded merganser	N						2
Common merganser	N						2
Turkey vulture	PN						2
Eastern screech-owl	PNR			2			
Barred owl	PN						2
N. saw-whet owl	PNR		2	1		2	
Northern flicker	PFNR			1		1	1
Pileated woodpecker	PFNR			1(>18)			
Yellow-bellied sapsucker	PFNR				1	1	1
Hairy woodpecker	PFNR		1	1	1	1	1
Downy woodpecker	PFNR	1	1		1	1	
Great crested flycatcher	N					2	2
Black-capped chickadee	NR	1					
Boreal chickadee	NR	1					
Tufted titmouse	NR	1					2
Red-breasted nuthatch	FNR				2	2	2
White-breasted nuthatch	FNR				2	2	2
Brown creeper	FNR		2	2	2	2	2
House wren	FNR	2			2	2	
Winter wren	N	2					
Eastern bluebird	FNR	2	2	2			
European starling	NR		2	2			2

[a] P—Perching   [b] DS—Dead, soft   [c] 1—Primary-cavity excavator
F—Foraging   DH—Dead, hard   2—Secondary-cavity user
N—Nesting   LCD—Live, central decay
R—Roosting   LBTL—Live, broken tops and limbs

mergansers, and pileated woodpeckers in New England (DeGraaf 1984). Additional foraging sites and small-diameter nesting/roosting sites can be expected regularly in tree crowns from natural mortality in northern hardwood stands.

Tree species longevity, rates and characteristics of decay, and lumber value affect the suitability and availability of cavity trees over time. Where choices of cavity trees are available, sugar maple is the most long-lived species that has good cavity potential, followed in decreasing order of longevity by American beech, white ash, red oak, red maple, and aspen.

The capacity to produce large trees varies among northern hardwood stands. Very good hardwood sites commonly produce few conifers and oaks. The time between harvests (rotations or cutting cycles) affects the size of trees in the stand and their species composition. Rotations of 60 years or less on average sites produce few large trees. Rotations of 65 to 100 years produce some smaller diameter cavity trees. These are probably adequate for most cavity-dwelling species but not for the larger species. Rotations longer than 100 to 125 years can be expected to produce cavity trees for larger cavity-using species. Longer cutting cycles or periods between subsequent stand entries (>15 years) increase the availability of cavities in managed stands (Tubbs et al. 1987).

LANDSCAPE CONSIDERATIONS

The changes in stand development and, therefore, breeding bird composition that take place in northern hardwood types also occur in other forest cover types that contain different birds. The landscape is a mosaic of forest types, successional stages, wetlands, and various open habitats, all of which contribute to the regional diversity of wildlife. Farming is declining, and old-field succession continues to produce changes in New England's wildlife habitats.

Local ecosystems may change dramatically as a result of land-use changes and succession; timber management changes bird habitats in a short time, as the example using northern hardwoods has shown. The landscape is a mosaic of stands and local ecosystems, and knowledge of the changes that can or will occur is useful for planning the continued availability of bird habitats in particular and wildlife diversity in general.

Glaciation produced profound changes in the New England landscape, and the manner in which the transported debris or drift was deposited largely determines soil fertility. Because soils did not develop in place, New England is not simply a region of infertile uplands grading into fertile valleys. Often the best soils for forest development consist of unsorted glacial drift (till) deposited during glaciation. Such soils are frequently found on midslopes of hills and mountains, and they produce northern hardwood forests of various compositions. Conversely, water-transported drift, frequently stratified and deposited in broad outwash plains and glacial lake shores, often produced poor soils at lower elevations. Such outwash areas typically produce aspen or gray birch barrens or oak woodlands (Barrett 1962).

Till soils throughout New England frequently contain a relatively impervious compact layer, or fragipan, several (10 cm to 30 cm) or more below the surface. Vernal pools, seeps, and wet ground, even on upper slopes, are common in spring because of the presence of this layer.

The forested New England landscape is a mosaic of forest types and nonforest habitats that occur in relatively small patches in most places, especially in the mountains.

Major forest cover types, in addition to northern hardwoods, include aspen, paper birch, red maple, northern red oak, white pine/red oak/red maple, balsam fir, white pine, red spruce/balsam fir, red spruce, and eastern hemlock. Pitch pine (*Pinus rigida*)-scrub oak (*Quercus illicifolia*) is common on Cape Cod, but otherwise it is of minor occurrence on some outwash plains in the region. The wildlife communities potentially associated with various timber size-classes of the major types and nonforest habitats have been compiled (DeGraaf and Rudis 1986). There are more than 330 vertebrates distributed among the inland habitats of New England; some are fairly ubiquitous, whereas others are quite habitat-specific.

Some species occur only in forested habitats; others occur only in coniferous forest (Cape May and pine warblers, for example); and still others occur only in deciduous forest (ovenbird, red-eyed vireo). Among species occurring only in nonforest habitats, least bitterns, teal, and marsh wrens occur only in wetlands, whereas meadowlarks occur only in upland fields (Table 8.3.). Others appear at the interfaces of forested and open habitats (e.g., indigo buntings and red-tailed hawks).

We already know many of the structural elements of the landscape (i.e., how species are distributed by habitat type and how habitats change with succession and stand development). We know somewhat less about functional relationships, such as movements of species between ecosystems. Once these are known, models can shift their emphases from managed timber stands and local ecosystems to the landscape as a whole. After unusual landscape features and habitats for rare species have been re-

TABLE 8.3.
Habitat Specificity of Some New England Breeding Birds
(after DeGraaf and Rudis 1986)

Species	Landscape Feature	Specific Habitat	Key Requirement
American robin	Forest		
Dark-eyed junco	Forest		
White-throated sparrow	Forest		
Wood thrush		Mature forest	
Magnolia warbler		Conifer forest	
Pine warbler		Mature conifer forest	
Yellow-throated vireo		Mature deciduous forest	
Hairy woodpecker		Mature forest	Cavity tree
Olive-sided flycatcher	Forest	Recent clearcut	High perch, wet spot
American kestrel	Open		Cavity tree
Northern flicker	Semiopen		Cavity tree
Field sparrow	Old field		
Eastern meadowlark	Open/semi-open	Grassy field	
Red-tailed hawk	Semiopen	Field/forest ecotone	
Savannah sparrow	Open	Upland field	
Common loon	Lake		Freedom from disturbance
Pied-billed grebe	Wetland	Shallow marsh	
Green-winged teal	Wetland	Shallow marsh	
Sedge wren	Wetland	Sedge meadow	
Great blue heron	Wetland	Shallow marsh	Tall dead trees
Common merganser	Stream/river	Wooded shoreline	Cavity trees

served, the remainder of the landscape can be managed to meet the habitat needs of all potential species. Only at the landscape scale can habitats for all species be provided simultaneously and over time.

Literature Cited

Aber, J. D. 1979. Foliage height profiles and succession in northern hardwood forests. Ecology 60:18–23.

Absalom, S. 1988. Comparison of avian community structure and habitat structure in mature versus old-growth northern forests. M.S. Thesis, Univ. Massachusetts, Amherst. 80pp.

Anderson, S. H., and H. H. Shugart. 1974. Habitat selection of breeding birds in an east Tennessee deciduous forest. Ecology 55:828–837.

Barrett, J. W. 1962. Regional silviculture of the United States. Ronald Press, NY. 610pp.

Bormann, H. F., and G. E. Likens. 1979. Pattern and process in a forested ecosystem. Springer-Verlag, NY. 253pp.

Carey, A. B., and W. M. Healy. 1981. Cavities in trees around spring seeps in the maple-beech-birch forest type. U.S.D.A. For. Serv. Res. Pap. NE–480.

Cooley, J. H. 1964. The effect of selection cutting on cull in northern hardwoods. J. Forestry. 62:823–824.

DeGraaf, R. M. 1984. Managing New England woodlands for wildlife that uses tree cavities. Bull. C-171. Univ. Mass. Coop. Ext. Serv., Amherst, MA. 17pp.

————. 1987. Managing northern hardwoods for breeding birds. Pages 348–362 *in* R. D. Nyland, ed. Managing northern hardwoods. Misc. Publ. 13, SUNY Coll. Environ. Sci. and For., Syracuse, NY.

DeGraaf, R. M., T. W. Anderson, and E. H. Zube. 1976. Relating wildlife to scenic resource value. Man-Environment 6:63–64.

DeGraaf, R. M., and N. L. Chadwick. 1987. Forest type, timber size class, and New England breeding birds. J. Wildl. Manage. 51:212–217.

DeGraaf, R. M., and D. D. Rudis. 1986. New England Wildlife: habitat, natural history, and distribution. Northeast. For. Exp. Sta., Gen. Tech. Rep. NE-108. U.S.D.A. For. Serv., Broomall, PA. 491pp.

Filip, S. M. 1978. Impact of beech bark disease on uneven-aged management of a northern hardwood forest. Northeast. For. Exp. Sta., Gen. Tech. Rep. NE-45. U.S.D.A. For. Serv., Broomall, PA. 7pp.

Goldstein, E. L., M. Gross, and R. M. DeGraaf. 1986. Breeding birds and vegetation: a quantitative assessment. Urban Ecology 9:377–385.

Gullion, G. W. 1984. Managing northern forests for wildlife. The Ruffed Grouse Soc., Coraopolis, PA. 72pp.

Healy, W. M. 1987. Habitat characteristics of uneven-aged stands. Pages 338–347 *in* R. D. Nyland, ed. Managing northern hardwoods. Misc. Publ. 13, SUNY Coll. Environ. Sci. and For., Syracuse, NY.

Healy, W. M., R. T. Brooks, and R. M. DeGraaf. 1989. Cavity trees in sawtimber-size oak stands in central Massachusetts. North. J. Applied For. 6:61–65.

Hilden, O. 1965. Habitat selection in birds. Ann. Zool. Fenn. 2:53–75.

Holmes, R. T., and S. K. Robinson. 1981. Tree species preferences of foraging insectivorous birds in a northern hardwoods forest. Oecologia 48:31–35.

James, F. C. 1971. Ordinations of habitat relationships among breeding birds. Wilson Bull. 83:215–236.

Karr, J. R. 1968. Habitat and avian diversity on strip-mined land in east-central Illinois. Condor 70:348–357.

Karr, J. R., and R. R. Roth. 1971. Vegetation structure and avian diversity in several new world areas. Am. Natur. 105:423–435.

Lack, D. 1933. Habitat selection in birds. J. Animal Ecol. 2:239–262.

Leak, W. B. 1964. An expression of diameter distribution for unbalanced, uneven-aged stands and forests. For. Sci. 10:39–50.

Leak, W. B. 1965. The J-shaped probability distribution. For. Sci. 11:405–409.

Leak, W. B. 1979. Effect of habitat on stand productivity in the White Mountains of New Hampshire. U.S.D.A. For. Serv. Res. Pap. NE–452. Broomall, PA. 8pp.

MacArthur, R. H., and J. W. MacArthur. 1961. On bird species diversity. Ecology 42:594–598.

Noon, B. R., and K. P. Able. 1978. A comparison of avian community structure in the northern and southern Appalachian Mountains. Pages 98–117 in R. M. DeGraaf, tech. coord. Management of southern forests for nongame birds. U.S.D.A. For. Serv. Gen. Tech. Rep. SE–14. Asheville, NC.

Roth, R. R. 1976. Spatial heterogeneity and bird species diversity. Ecology 57:773–782.

Trimble, G. R., Jr. 1963. Cull development under all-aged management of hardwood stands. U.S.D.A. For. Serv. Res. Pap. NE–10, Upper Darby, PA. 10pp.

Tubbs, C. H. 1977. Manager's handbook for northern hardwoods in the North Central states. U.S.D.A. For. Serv. Gen. Tech. Rep. NC–39, St. Paul, MN. 29pp.

Tubbs, C. H., R. M. DeGraaf, M. Yamasaki, and W. M. Healy. 1987. Guide to wildlife tree management in New England northern hardwoods. U.S.D.A. For. Serv. Gen. Tech. Rep. NE–118. Broomall, PA. 30pp.

Webb, W. L., D. F. Behrend, and B. Saisorn. 1977. Effect of logging on songbird populations in a northern hardwood forest. Wildl. Monogr. No. 55. The Wildlife Society, Washington, DC. 35pp.

Willson, M. F. 1974. Avian community organization and habitat structure. Ecology 55:1017–1029.

APPENDIX 8.1.
Scientific names of birds occurring in New England northern hardwoods.

Common loon (*Gavia immer*)
Pied-billed grebe (*Podilymbus podiceps*)
Least bittern (*Ixobrychus exilis*)
Great blue heron (*Ardea herodias*)
Wood duck (*Aix sponsa*)
Green-winged teal (*Anas crecca*)
Common goldeneye (*Bucephala clangula*)
Hooded merganser (*Lophodytes cucullatus*)
Common merganser (*Mergus merganser*)
Turkey vulture (*Cathartes aura*)
Red-tailed hawk (*Buteo jamaicensis*)
American kestrel (*Falco sparverius*)
Ruffed grouse (*Bonasa umbellus*)
Eastern screech-owl (*Otus asio*)

Barred owl (*Strix varia*)
Northern saw-whet owl (*Aegolius acadicus*)
Yellow-bellied sapsucker (*Sphyrapicus varius*)
Downy woodpecker (*Picoides pubescens*)
Hairy woodpecker (*Picoides villosus*)
Northern flicker (*Colaptes auratus*)
Pileated woodpecker (*Dryocopus pileatus*)
Olive-sided flycatcher (*Contopus borealis*)
Willow flycatcher (*Empidonax traillii*)
Least flycatcher (*Empidonax minimus*)
Great crested flycatcher (*Myiarchus crinitus*)
Tree swallow (*Tachycineta bicolor*)

Black-capped chickadee (*Parus atricapillus*)
Boreal chickadee (*Parus hudsonicus*)
Tufted titmouse (*Parus bicolor*)
Red-breasted nuthatch (*Sitta canadensis*)
White-breasted nuthatch (*Sitta carolinensis*)
Brown creeper (*Certhia americana*)
House wren (*Troglodytes aedon*)
Winter wren (*Troglodytes troglodytes*)
Sedge wren (*Cistothorus platensis*)
Marsh wren (*Cistothorus palustris*)
Golden-crowned kinglet (*Regulus satrapa*)
Eastern bluebird (*Sialia sialis*)
Veery (*Catharus fuscescens*)
Swainson's thrush (*Catharus ustulatus*)
Wood thrush (*Hylocichla mustelina*)
American robin (*Turdus migratorius*)
Cedar waxwing (*Bombycilla cedrorum*)
European starling (*Sturnus vulgaris*)
Solitary vireo (*Vireo solitarius*)
Yellow-throated vireo (*Vireo flavifrons*)
Philadelphia vireo (*Vireo philadelphicus*)
Red-eyed vireo (*Vireo olivaceus*)
Northern parula (*Parula americana*)
Chestnut-sided warbler (*Dendroica pensylvanica*)
Magnolia warbler (*Dendroica magnolia*)
Cape May warbler (*Dendroica tigrina*)

Black-throated blue warbler (*Dendroica caerulescens*)
Black-throated green warbler (*Dendroica virens*)
Blackburnian warbler (*Dendroica fusca*)
Pine warbler (*Dendroica pinus*)
Black-and-white warbler (*Mniotilta varia*)
American redstart (*Setophaga ruticilla*)
Ovenbird (*Seiurus aurocapillus*)
Mourning warbler (*Oporornis philadelphia*)
Common yellowthroat (*Geothlypis trichas*)
Canada warbler (*Wilsonia canadensis*)
Scarlet tanager (*Piranga olivacea*)
Rose-breasted grosbeak (*Pheucticus ludovicianus*)
Indigo bunting (*Passerina cyanea*)
Rufous-sided towhee (*Pipilo erythrophthalmus*)
Field sparrow (*Spizella pusilla*)
Savannah sparrow (*Passerculus sandwichensis*)
White-throated sparrow (*Zonotrichia albicollis*)
Dark-eyed junco (*Junco hyemalis*)
Eastern meadowlark (*Sturnella magna*)
American goldfinch (*Carduelis tristis*)

9

Wildlife Communities of Southwestern Riparian Ecosystems

ROBERT C. SZARO

THERE ARE landscapes within any ecosystem that assume a value disproportionate to their areal extent. The southwestern riparian ecosystems of Arizona and New Mexico are typical examples of such landscapes. The vegetation that grows along surface and subsurface waterways forms a unique and threatened ecosystem in need of new management considerations. Historically, these stream and riparian landscapes have been of substantial economic importance to grazing, farming, mining, recreational opportunities, and residential developments.

The region's population growth over the last two decades has exerted increased pressure on these rapidly disappearing landscapes. Recently, federal and state agencies have acted to protect, preserve, and develop guidelines for the proper management of riparian zones. A major contributor to this effort is Dr. Szaro, working out of the U.S.D.A. Forest Service in Tempe, Arizona.

For the past decade, Szaro has defined the structure, function, and values of stream and riparian ecosystems in the southwestern United States. Riparian protection is a unique American concept. Its roots are anchored in the work begun by the U.S. Fish and Wildlife Service's Natural Wetland Inventory of Wetland and Aquatic Habitats program. The concept of riparian and stream protection is now beginning to take shape.

Szaro emphasizes the need to recognize that the more rapidly the landscape changes from over use the more difficult it will be for ecosystems and human activities to adapt to changes or losses.

Local land users originally feared protective zoning of portions of these landscapes. To some extent this fear has subsided because the land users realized they will not be deprived of the resource values they desire in the long run. Protective zoning in carefully selected locations demonstrates not a hindrance but rather a concern to promote an economic base for continued grazing interests and recreational and residential developments.

174

INTRODUCTION

Southwestern riparian ecosystems are disproportionately rich in bird and other wildlife, but because these systems are linear and often represent a small portion of the landscape, they frequently have been overlooked in management prescriptions. For example, riparian areas comprise an estimated 1.3% of the total lands within the boundaries of national forests in Arizona and New Mexico (Szaro 1989). Because small scales previously were used in planning and managing applications, riparian systems typically were included within surrounding vegetation types. But increasing recognition of the importance of these areas has resulted in a new emphasis regarding their management. Management of these areas poses formidable problems for land managers.

Riparian communities in the Southwest traverse the entire range of life zones from alpine communities of the highest mountains to subtropical Sonoran Desert scrub plains and valleys of the lower Gila and Colorado rivers. They are composed both of distinctive riparian species and of denser and larger specimens of species common elsewhere. A key feature of riparian environments is that these systems are in constant flux. Adaptations influence the development of riparian ecosystems where change is the rule, and a true steady state probably never exists (Heede 1985). The dynamics of riparian plant communities present a complex set of interacting variables affected by both time and environment. Superimposed on the equilibrium fluctuating between the physical environment and plant species is the animal species component, which utilizes these areas as breeding and wintering habitat, migratory corridors, and as areas for brief stopovers. Wildlife communities associated with southwestern riparian systems also are affected by the vegetative communities of the surrounding uplands. Faced with large species complexes and the patchy nature of many riparian corridors, land managers look for indicators of habitat health and condition. Mobile species, such as birds, can have broad elevational and habitat ranges, whereas more sedentary species, such as small mammals, reptiles, and amphibians, are much more limited. Further complicating any riparian management plans is the fact that these same areas are prime sites for human occupancy and use, and much biologically productive bottomland is now in private ownership. Many riparian forests have been cleared and the lands now farmed or intensively grazed (Kauffman and Krueger 1984). Nonfarm homesites also occur throughout much of the habitat type. Grazing, recreation, timber harvest, water enhancement and control projects, and naturally occurring disturbances all combine to make riparian systems difficult to manage.

Although riparian animal species may not be responding directly to differences in abiotic site characteristics, they certainly are closely tied to their habitats. Because riparian plant communities respond to differences in elevation, stream gradient, and stream direction, these factors also indirectly influence their distribution and abundance. At a particular elevational zone, riparian animal species are heavily influenced by the structure and floristic composition of their habitats. For some species, particularly birds, structure is most important. For others, such as lizards, structure alone does not guarantee their presence; these species are more closely tied to plant species composition.

Successful management of southwestern riparian communities requires a suite of strategies to enhance and maintain riparian diversity. An ecosystem approach to management of riparian systems should consider goals and standards for the condition of major indicators of plant and animal community health. It should provide for the continued monitoring of habitat indicators and of species' population levels over an extended time period (5+ years) to fully assess the effects of environmental perturbations and the adequacy of management strategies. Several studies in the southwestern United States that are beginning, in progress, or nearing completion address the need for the long-term ecosystem approach. Conceivably, results from these and future such studies will be used by land-management agencies to more effectively manage riparian areas. In this paper I discuss appropriate management actions, factors affecting riparian-zone management, and examples from case studies that support the concept of ecosystem management. Finally, I propose recommendations for future research and management of riparian areas in Arizona and New Mexico.

DETERMINING APPROPRIATE MANAGEMENT ACTIONS

Biologists are asked to assess the impact of some proposed management action, often with little or no site data on hand. This unfortunate scenario is unlikely to change in the future as society rapidly expands and uses more and more of the shrinking natural resource base. It will be further exacerbated as funding and manpower constraints preclude the possibility of long-term monitoring and assessment. Yet, timely management decisions need to be made if we hope to have any chance in mitigating environmental impacts. Usually, a land manager is called upon to evaluate the health of a biological system by assessing its degradation from any of a variety of man-caused impacts (Karr 1987). The decision-making process requires using all available techniques. Documenting the relationships between vertebrates and physical or biotic features of their habitat is important in preserving adequate numbers of individuals or species (Rotenberry 1985). If used properly, biological data can effectively assist the land manager in assessing potential impacts of proposed habitat-management activities and in identifying management opportunities for wildlife species (Verner and Boss 1980). Even if we sometimes make the wrong choices, it is better to try than simply to throw up our hands and say we do not have the data. Unfortunately, our ability to monitor trends in the abundance of animals in response to management practices is short of the technical development needed to do the job properly and efficiently (Verner 1986). The establishment of consistent, repeatable relationships between an animal and components of its habitat permits the prediction of species abundance as a function of those components (Rotenberry 1985). But, how do we go about determining these relationships with those data we have and/or can afford to collect?

Basic Steps in the Decision-making Process

We can greatly increase the odds of making environmentally sound management decisions by following several steps that take advantage of all the information that is currently available, allowing us to make an informed best guess. As we add more information to the process, the confidence and accuracy of our recommendations will increase.

DETERMINING SCOPE AND SCALE

First, the manager needs to ask some preliminary questions about the scope and scale of any proposed land-management activity and the resolution of the data. This includes identifying issues, concepts, or opportunities in the initial problem definition. Generalities should be avoided and the emphasis placed on specifics and prioritizing objectives. *Is the problem at the national, regional, local, or microhabitat level of concern?* The smaller the project site, the greater the need for biological expertise in refining the wildlife predictions (Verner and Boss 1980). The data resolution necessary for broad management decisions for large areas is potentially less than that needed for smaller project sites because large areas tend to have greater heterogeneity and interspersion of habitat types and seral stages.

DATA COLLECTION AND AVAILABILITY

The data should be tailored to fit the project area. Literature searches, previous assessments from similar areas, museum reviews, and consultations with other area biologists and university personnel can provide lists of species of concern. Managers should take advantage of all available data sources before assessing the necessity of any additional fieldwork. If a need to conduct thorough field investigations is determined, the biologist should use preliminary data as a framework for further fieldwork.

DATA ANALYSIS, EVALUATION, AND INTERPRETATION

The hardest step in any environmental analysis is the assimilation and organization of all the available and frequently conflicting data. Many techniques are used to determine the relationships between animals and their habitats. Once the basic habitats and broad wildlife effects have been determined, the biologist will need to refine the analysis based on such things as interspersion of habitat types, the presence or absence of special habitats, specific requirements of species for water, space, and cover, and a host of other factors beyond broad habitat needs that enter into determining habitat suitability for a particular species (Verner and Boss 1980). Techniques used for determining habitat relationships range from simple correlation analysis to advanced multivariate methods, such as cluster analysis, discriminant function analysis, principal component analysis, and multiple regression. The result of many of these analyses then can be used in the construction of a predictive model to aid in the decision-making process. But where do we start, and once we complete the analysis, how do we incorporate this information into a predictive or decision-making tool?

Assessments of the relationships between animal community structure and population levels with the physical configuration of the environment over broad biogeographic ranges have demonstrated strong patterns of association. Similar associations on a more limited local scale require more site-specific information to develop. Determining these relationships and the potential effects of changing or modifying important habitat parameters are keys to our understanding and assessment of proposed management actions. An assortment of techniques is available to assist us in determining these relationships. None is the best method, but some do have a greater degree of reliability. The underlying premise to all these techniques is that we have adequate and reliable data available. However, depending on the scope and scale of any proposed management action, lesser or greater amounts of data are needed for a sound decision.

FACTORS AFFECTING RIPARIAN-ZONE MANAGEMENT

Changes in vegetative structure that result from both natural and human-related disturbances are known to affect dramatically the richness and densities of riparian bird species (Szaro 1980, Szaro and Debano 1985). Typical disturbances include flooding, inundation, scouring, desiccation, grazing, recreation, dam construction, and revegetation (Hendrickson and Minckley 1984, Minckley and Brown 1982). The effects of disturbance on biological systems are often inferred from changes in species diversity, stability, and biomass (Cole 1987). Habitat management decisions for wildlife frequently are made by considering a limited set of species (e.g., common game species or endangered species). Some groups, such as birds, respond quickly to changes in habitat structure because of their mobility. By contrast, small mammals, reptiles, and amphibians are more limited in their ability to immigrate or emigrate to or from an impacted area. Similarly, fish species are limited to movements within their stream system. Unfortunately, few studies of riparian fauna have examined all vertebrate groups and their relationships with structural habitat components, vegetative species composition, and both natural and human-related disturbances. This lack of research could lead to erroneous assumptions about potential effects of overall resource management practices on wildlife, resulting in undesired positive or negative effects on other vertebrate species.

Floristic Versus Structural Diversity

How valid are arguments against distinguishing extremely similar plant species when determining animal species relationships? Is a detailed analysis based on species composition versus those based on growth form or even genera necessary? Is it important to a manager to know there are 2, 5, 10, or even 20 kinds of willow (*Salix* spp.) communities, separable only by experts or by considerable effort with plant keys, as opposed to tall and short willow communities? The answers to these questions pose some fundamental problems in ecological research and especially to the relationships between animal faunas and riparian plant communities.

Studies on birds have demonstrated strong patterns of association between bird community structure and vegetation structure and habitat configuration (MacArthur and MacArthur 1961). Additional studies have emphasized that habitat physiognomy is more important than the particular floristics of the vegetation (Wiens and Rotenberry 1981). However, subsequent investigations, on a more local scale, have discovered strong correlations between avian community composition and shrub species composition (Wiens and Rotenberry 1981). The use of summary indices that ignore floristic information likely has impeded detection of relationships between bird species and plant taxa (Rotenberry 1985). Increasing numbers of studies demonstrate significant associations between individual bird species and particular plant species at the local level. This result is hardly unique to birds; however, as Szaro and Belfit (1986) found that composite variables that mask taxonomic composition of plant species were not good predictors of reptile and amphibian abundance behind Whitlow Ranch Dam along Queen Creek, Arizona, whereas densities of specific plant species were good predictors of abundance. Rotenberry (1985) suggested that such associations that likely exist in most systems are obscured by loss of information that results from summarizing community composition and structure. Composition and structure are extremely complex and intrinsically multivariate in nature and hard to reduce into a few simple indices.

Clearly then, no simple answers exist to the questions proposed above, in spite of the desire for simplification. Not enough work has been done to ascertain whether groupings such as *tall* or *short* willow communities make ecological sense. I suspect, as more and more relationships are examined in detail, the need for specific floristic information will become more apparent. Furthermore, it is always possible to group communities later, such as a high-elevation scrub-willow type, if types are determined on a floristic basis. The reverse is not true. The level of detail should be determined by the goals and needs of the individual or agency. Much ecological work has been plagued by oversimplification. All available floristic information must be carefully analyzed in order to better understand and manage riparian systems.

The Role of Disturbance in Riparian Systems

Disturbance plays an integral role in establishment and development of southwestern riparian ecosystems. It occurs frequently enough in many systems to allow the coexistence of species with varying degrees of competitive ability (Pickett 1980). Periods of standstill in the disturbance process, called dynamic equilibrium, alternate with periods of change (Heede 1985). Changes result from periodic, abrupt, and/or catastrophic environmental factors, leading to displacement, replacement, and succession, with species composition being a function of a disturbance regime (White 1979).

Frequent disruption of the community and creation of open sites often results in species mixtures that are fleeting in time and do not repeat in space (White 1979). Succession is a diverse process; it is based on invasion, initial floristic composition, life span, tolerance, and levels of disturbance differing in frequency, predictability, and magnitude (Pickett 1980). The lack of continued recruitment of important canopy species indicates that over time, as existing adult trees age and die, these systems can experience large structural and floristic shifts. This is particularly true of riparian ecosystems, which often are characterized by being remarkably distinct and highly integrated intermittent pockets within other communities. For example, high-elevation species often extend downslope within canyons that lead cooler and moister air downward (Minckley and Brown 1982).

GRAZING

One of the most prevalent forms of disturbance in riparian zones is heavy livestock use (Johnson and Jones 1977). Grazing is so ubiquitous in riparian ecosystems of the Southwest that only a few ungrazed sites are available for study as references or controls. Stream bottoms are natural concentration areas for livestock (Platts and Nelson 1985), and in arid or semiarid rangelands this attraction is intensified by limited water on the drier adjacent uplands.

Degradation of many riparian systems and overgrazing by domestic livestock are thought to be related integrally (Hendrickson and Minckley 1984, Platts and Wagstaff 1984). Grazing exerts a strong influence on riparian soils, vegetation, and animal communities (Cannon and Knopf 1984, Jones 1981). Livestock may be the major cause of excessive habitat disturbance in most western riparian communities (Mosconi and Hutto 1982, Taylor 1986). Prime characteristics of riparian ecosystems that are conducive to heavy livestock use include plentiful amounts of succulent forage, moderate slopes allowing easy accessibility, shade, a generally reliable water supply, and a microclimate more favorable to cattle than that of surrounding terrain (Kauffman and Kreuger 1984). More-or-less unrestricted livestock use dates back to the earliest days of settlement by Caucasians, and before (Johnson and Jones 1977).

Grazing impacts riparian vegetation in two ways—removal and trampling (Kauffman and Krueger 1984). Removal of vegetation potentially reduces both numbers and biomass of plant species. Vegetation is physically damaged by rubbing, trampling, and browsing. Moreover, grazing and/or browsing can alter growth form of plants by removing terminal buds and stimulating lateral branching. Grazing also influences the spacing of plants, width of the riparian corridor, seedling establishment, and species composition (Mosconi and Hutto 1982). Consequently, floristic diversity is often lower in grazed areas.

Improper livestock use not only directly affects riparian vegetation, but also the physical habitat. Trampling increases soil compaction, contributes to stream-bank erosion, decreases water quality, widens and shallows channels, and physically destroys vegetation (Kauffman and Krueger 1984). Grazing is so ubiquitous on riparian ecosystems of the Southwest that few ungrazed examples remain. Ungrazed sites exist only in those unique situations where topography, exclosures on public lands, or exclusion from private holdings limits livestock accessibility.

Fencing and complete exclusion of grazing from riparian ecosystems often is identified as the only livestock-grazing strategy capable of rehabilitating damaged stream courses (Platts and Wagstaff 1984). Many studies have reported adverse effects of cattle grazing on riparian vegetation and of recovery of riparian vegetation when grazing is modified, reduced, or eliminated (Cannon and Knopf 1984, Taylor 1986). Although fencing and livestock exclusion would certainly alleviate part of the problem, it may not be economically feasible on a broad scale. Light to moderate late-season grazing has been suggested as one possible alternative (Kauffman and Krueger 1984). Martin (1979) recommended early-season grazing and removal of animals before or soon after the more palatable tree species put on leaves, as tree sprouts become more attractive after they leaf out. Further research is needed on livestock distribution practices, as well as on some innovative new ideas, such as behavioral modifications of cattle herds (Severson, pers. comm.), before a solution to this problem is found.

FLOODING

Flooding is the most common naturally occurring form of disturbance in riparian ecosystems. Flood frequency, flow duration, and period of inundation are independent hydrologic factors known to affect vegetation patterns (Hupp and Osterkamp 1985). Shifting patterns of vegetation, in which species of varying colonizing ability, tolerance to flooding, and shade tolerance locate along disturbance-created gradients, result from stream migration, erosion, and deposition (White 1979). The natural processes involved when changes occur in floodplains are complex and varied, leading to spatial and temporal shifts in stream channel width, length, and sinuosity, as well as the areal extent of vegetation.

Flooding effects can be destructive or constructive to riparian plant communities, depending on flow volume and velocity. The destructive effects of flooding are relative, and all riparian ecosystems can experience flood effects that are magnified by other disturbances (e.g., grazing, timber harvest, and recreation) beyond their ability to recover. Storm flows have greater force than normal flows, raise the water level, usually increase turbulence, and are more likely to have eroding velocities. Major storm flows exert great force on stream-channel banks and on objects in the main-flow path. These forces are often of such magnitude that debris and vegetation on the floodplain are flushed downstream, and the stream channel is altered extensively. Stresses produced by the turbulent forces along stream banks and around other stationary objects cause bank erosion, tree uprooting, and destruction of alluvial fans, resulting in additional debris in the flowing water (Burkham 1972). Consequently, floods remove and drown plants in addition to scouring stream channels and root systems. Proximity to the stream course, duration of inundation, frequency and intensity of flooding, substrate type, and plant height (age) all determine susceptibility of riparian plants to flooding (Hupp 1982). In contrast, floods also deposit sediments and provide moist conditions for seedling germination and establishment (Fenner et al. 1985). Vegetation and flow are normally in equilibrium, the plant species present in any one reach being those that can tolerate both normal and storm flows of the reach.

Many riparian plant communities are dependent on light to moderate flooding for successful reproduction (Brady et al. 1985). Reproductive strategies and other ecological characteristics of riparian trees are strikingly adjusted to a disturbance regime (White 1979). Cottonwood (*Populus* spp.), willow, and seepwillow (*Baccharis* spp.) seeds germinate quickly, either on water or moist soil, and require wet soils for several weeks for establishment (McLeod and McPherson 1973). Because of reduced water velocities, secondary channels are often completely vegetated, whereas mainstream courses are kept barren by scouring at times of high flow (Brady et al. 1985). Moreover, abundant streamside vegetation can further minimize impacts of flooding (Platts et al. 1985). This vegetation often acts as a barrier to the passage of organic debris during floods, creating piles of accumulated materials. These flood-caused heaps are important features for small mammals, amphibian, and reptile communities (see Szaro et al. 1988a for an example).

Presence of certain riparian species may be an integral part of a recruitment process. Seepwillow, arrow-weed (*Tessaria sericea*), and other woody shrubs have remarkable capacities to resist scouring by floods (Minckley and Clark 1984). Accrual of sediments is enhanced by reduced flow velocity and volume through stands of these plants and the trapping and accumulation of organic debris (Minckley and Rinne 1985). This creates a substrate favorable to seedling establishment, and successful stand occurrence largely depends on favorable flood volumes during stand development (Brady et al. 1985).

Regulation and Damming of Streams

Floristic and structural diversity along stream courses are usually both visually and ecologically desirable, but riparian wildlife is often impoverished as a result of river-channel modifications carried out to alleviate flooding and improve land drainage. Functionally, riparian habitats along unregulated streams and rivers are open ecosystems, which are mutually related to their adjacent environments and are characterized by a lateral flow of water together with seasonal fluctuations in water level (Nilsson 1982). Under natural conditions, flood discharges periodically submerge portions of the riverbank and/or floodplain, thereby rearranging sedimentary deposits (Petts 1984). Any disruption of normal flow patterns results in changes in development of riparian communities.

Dams and water diversions are known to change significantly downstream flow regimes, levels of winter floodwaters, dry-season flow rates, and riparian-zone soil moisture. The loss of pulse-stimulated responses at the water-land interface is an important downstream effect of river impoundment. Changes upstream from a dam, such as water impoundment and salt accumulation, are even more striking (Turner and Karpiscak 1980). Native riparian plants are generally unsuccessful in recolonizing the new shores because the natural, seasonal waterlevel variation is almost completely reversed in reservoirs, compared to that of the lakes and rivers before regulation. The straightening, widening, and deepening of river channels during major engineering works usually reduces both in-channel and riparian habitat diversity (Nilsson 1982).

Fundamental differences between natural habitat conditions and those of reservoirs and regulated streams and rivers lead to a whole spectrum of changes that are much more complex than previously thought (Nilsson 1982). In reservoirs, waterlevels may be kept high for long periods of time, whereas in streams and rivers long periods of high-waterlevels often do not occur. In fact, low flows and periods of no flow frequently occur under normal regulation regimes, with the pattern of controlled releases almost certainly distributed differently than those of the natural flow regime. Drastic changes in waterlevel, as in reservoirs, can lead to the destruction of natural vegetation, although some species may be able to recolonize the new shoreline.

The timing and duration of flooding also are critical, having greater impact during the growing season than during the dormant season, when flooding may not adversely affect tree growth and may even stimulate it (Kozlowski 1984). Mortality varies with duration of flooding, species, and age of trees. Flood tolerance of older trees is generally greater than that of seedlings or saplings of the same species. In time, riparian vegetation can be expected to regain its dynamic equilibrium and become adjusted to changed waterlevel regimes, with modifications in both structure and composition (Nilsson 1982). Under conditions of constant flow, riparian vegetation often becomes dominated by trees, although in the southwestern United States, exotic saltcedar (*Tamarix pentandra*) frequently becomes the successful colonizer (Turner and Karpiscak 1980). Regulated flows favoring development of saltcedar on the Salt River,

Arizona, also substantially changed the habitat in ways unfavorable to cottonwood regeneration (Fenner et al. 1985). There was less floodplain inundation, no normally occurring large winter/spring floods, and consequently no creation of alluvial seedbeds. Channel incision further lowered water tables adjacent to the river, precluding seedling establishment because of an increased depth to sustained groundwater. Riparian vegetation plays an important role in determining the magnitude and characteristics of flood events, and the effectiveness of high flows for erosion within river channels has emphasized the role of riparian vegetation in channel change (Petts 1984).

RECREATION

Recreationists can and do damage riparian ecosystems, once carrying capacity has been reached and/or exceeded (Johnson and Carothers 1982, Manning 1979). Some of the same factors that attract livestock to riparian areas stimulate human activity (i.e., lush vegetation, shade, water, a generally more favorable microclimate, and an abundance of wildlife). Human activities that impact riparian resources include birdwatching, hiking, fishing, camping, hunting, trapping, picnicking, floating, boating, and river running (Johnson and Carothers 1982). As the amount of leisure time, personal income, mobility, and population levels have increased in the western United States, there have been corresponding increases in recreational effects on riparian sites (Johnson and Carothers 1982). Moreover, the recreational use per unit area of the riparian zone is many times greater than that of other vegetative types (Thomas et al. 1979). Human encroachment and activities have reduced the relative extent of these fragile ecosystems, making those that remain even more significant.

Effects of Site Characteristics on Riparian Plant Communities

Fluvial geomorphic processes determine the characteristics of a stream channel and its adjacent floodplain terraces. Stream gradient is an important parameter that influences most aspects of floodplain geomorphic and hydrologic processes (Hupp 1982). As water flows across the land surface in response to gravity, it loses potential energy as it loses elevation. When the amount of energy lost exceeds the gravitational forces acting upon materials in the streambed, sediments are picked up and deposited further downstream as stream channels change direction and gradients decrease.

The relationship between fluvial hydrologic processes and riparian vegetation patterns is not yet completely understood (Hupp 1982). However, vegetation patterns are related to topographic variation, with differences in elevation above the stream channel being a major factor determining species distributions (Hupp 1982, Hupp and Osterkamp 1985).

Elevation, stream bearing, stream gradient, and valley cross-sectional area are all factors that influence riparian plant communities. Altitudinal zonation of riparian communities has long been recognized (Minckley and Brown 1982), although absolute upper and lower limits may vary (Szaro 1989). Differences in elevation appear to be the overriding factor associated with distribution of riparian forest and scrub community types in Arizona and New Mexico (Szaro 1989). However, both stream gradient and stream bearing also may influence distribution but on a more limited and local level.

CASE STUDIES

Study Areas

Over the past 5 years, intensive studies have been conducted at riparian areas at upper, intermediate, and low elevations in the Southwest.

RIO DE LAS VACAS

The Rio de las Vacas is a montane stream draining the San Pedro Parks Wilderness Area, Santa Fe National Forest, New Mexico. The area is about 17 km southeast of the town of Cuba in Sandoval County at an elevation of 2600 m. Two cattle exclosures straddling the stream (both about 900 m long by 50 m wide) were installed in the early 1970s (Szaro et al. 1988a). Contiguous downstream private lands that are grazed by livestock were used for comparison with the nongrazed areas.

GARDEN CANYON

The riparian and adjacent communities were sampled along Garden Canyon, which is located on the Fort Huachuca Military Reservation near Sierra Vista, Cochise County, Arizona; elevations ranged from 1,500 m to 1,630 m. Riparian communities sampled, from lowest to highest elevation, were sycamore (*Platanus wrightii*), sycamore/juniper (*Juniperus monosperma*), and sycamore/juniper (*J. deppeana*)/oak (*Quercus arizonica, Q. emoryi,* and *Q. hypoluecoides*) (Szaro 1988). Plant communities sampled adjacent to the riparian corridor, from lowest to highest elevation, were composite (*Heterotheca* spp.)/grassland (*Poa* spp.), juniper (*J. monosperma*) woodland, and oak (*Q. emoryi*) woodland (Szaro et al. 1988*b*).

QUEEN CREEK

This study area is along Queen Creek in the Tonto National Forest, about 16 km west of Superior, Pinal County, Arizona, at an elevation of 625 m. An earth-fill dam constructed in autumn of 1960 has resulted in the dramatic development of an artificial riparian gallery forest of Goodding willow (*Salix gooddingii*) and saltcedar (*Tamarix pentandra*) that occupies an area of approximately 18 ha (Szaro and DeBano 1989).

Methods

AMPHIBIANS AND REPTILES

Different methods of capture were necessary for sampling the herpetological communities at the Rio de las Vacas versus Garden Canyon and Queen Creek. The most common reptile along the Rio de las Vacas is the wandering garter snake (*Thamnophis elegans vagrans*), which was not easily captured by pitfall traps. Consequently, snake populations were estimated using intensive searches by three to four observers for five consecutive days. Equal time was spent in each of three areas (upper exclosure, lower exclosure, and grazed stream segments) along the Rio de las Vacas. Time-of-day bias was minimized by alternating starting areas. Sampling begun in June 1985 was replicated in July, August, and September of that year and in the same months in 1986. All snakes were marked and released (Szaro et al. 1988*a*). In Garden Canyon, six trap stations were set in each of six habitats: composite/grassland, sycamore riparian, juniper woodland, sycamore/juniper riparian, oak woodland, and sycamore/juniper/oak riparian forest (36 stations in all). Trap stations consisted of two unbaited pitfalls (19 l or 5 gal) with a drift fence between buckets measuring 7.6 m long by 20 cm high. Pitfalls were checked three times each week from mid-April through the end of May and from mid-July through the first week of September 1985 and 1986. At Queen Creek, we established 40 pitfall trapping stations (20 each in the riparian and adjacent desert) similar to those used in Garden Canyon. Traps were checked three times a week during April and May 1982 (Szaro and Belfit 1986).

BIRDS

Birds were counted using variable–circular plots. At the Rio de las Vacas, birds were counted at 24 points (8 each in an upper exclosure, lower exclosure, and along the grazed stream segment), 6 times a month from June to September 1985 and 1986. At Queen Creek, birds were counted at 60 points (20 each in the riparian stand, desert wash, and desert upland), 10 times during the height of the breeding season, April to May 1981 and 1982 (Szaro and Jakle 1985).

SMALL MAMMALS

Along the Rio de las Vacas, small mammals were captured using Sherman live traps (8 by 9 by 23 cm) baited with rolled oats in a 40-trap grid pattern composed of 2 by 20 traps located parallel to the stream channel. Pairs of traps were located with one trap at the stream bank and the second 10 m from the bank. Trap pairs were separated from the next pair downstream by at least 5 m. Live traps were set for 5 nights each month from June to September 1985 and 1986; 40 each in the upper exclosure, lower exclosure, and in the grazed stream segments, for a total of 120 traps. Along Queen Creek and in Garden Canyon, small mammals were captured in the same pitfalls used for the amphibian and reptile study (Szaro and Belfit 1987, Szaro et al. 1988b).

Wildlife Responses

RIO DE LAS VACAS

At the high–elevation site, Rio de las Vacas, there was a broad spectrum of vegetative and vertebrate responses to exclosure of a 2-km section of the stream from cattle grazing. Bank stability along the Rio de las Vacas increased within exclosures compared to downstream grazed areas (Rinne 1988). Woody vegetation was markedly greater in the exclosures (Table 9.1.), as was stream-bank and overhanging woody vegetation in the exclosures.

There were differences in terrestrial vertebrates response to the exclusion of cattle grazing. First, in both 1985 and 1986, wandering garter

TABLE 9.1.

Estimates for Fish, Garter Snake, Bird, and Small Mammal Populations Along Rio de las Vacas, New Mexico

| | Treatment | | | | | |
| | Grazed | | Lower Exclosure | | Upper Exclosure | |
Species	1985	1986	1985	1986	1985	1986
Snakes						
Number of captures	20	10	164	120	115	179
Birds						
Number of observations	615	386	814	788	876	862
Number of species	36	34	44	44	44	47
Small Mammals						
Number of captures	52	25	189	31	138	30
Number of species	6	4	8	5	9	7

snake captures were markedly lower along grazed reaches of the Vacas than within exclosures (Table 9.1.). However, differences in snake captures between years were not consistent within exclosures. That is, captures decreased in the lower exclosure and increased in the upper exclosure from 1985 to 1986.

For birds, total individuals observed and species richness were markedly greater in ungrazed exclosures in both 1985 and 1986 than along the grazed stream reach (Table 9.1.). Species common in grazed areas were those whose foraging modes concentrated either on the ground or in the air. These species included Brewer's blackbird (*Euphagus carolinus*), northern flicker (*Colaptes auratus*), mountain bluebird (*Sialia currucoides*), pine siskin (*Carduelis pinus*), robin (*Turdus migratorius*), spotted sandpiper (*Actitis macularia*), violet-green swallow (*Tachycineta bicolor*), western bluebird (*S. mexicana*), western wood pewee (*Contopus sordidulus*), and yellow-rumped warbler (*Dendroica coronata*). In contrast, although all of these species were observed and some were as common or even more abundant in the exclosures, other species were more important components of bird-community structure. These included green-tailed towhee (*Pipilo chlorurus*), orange-crowned warbler (*Vermivora ruficapilla*), dark-eyed junco (*Junco hyemalis*), broad-tailed hummingbird (*Selasphorus platycercus*), and house wren (*Troglodytes aedon*).

Small mammal captures were markedly greater within ungrazed exclosures than along the grazed stream segment in 1985, but small mammal captures were similar in grazed and ungrazed reaches in 1986 (Table 9.1.).

Nevertheless, we captured fewer species on the grazed area than within either exclosure. Voles (*Microtus montanus*) and deer mice (*Peromyscus maniculatus*) were common in both grazed and exclosed areas, but chipmunks (*Eutamias minimus*) and golden mantled ground squirrels (*Spermophilus lateralis*) were common only in exclosures.

GARDEN CANYON

The effects of increasing structural and species complexity on amphibian, reptile, and small mammal populations and species richness were examined along an elevational gradient in Garden Canyon. Although amphibian and reptile species richness and total abundance were not significantly different (P > 0.05) between any of the 6 habitats examined, there were significant species differences (Table 9.2.). The ground-foraging whiptails (*Cnemidophorus sonorae* and *C. uniparens*) were very abundant in the structurally less complex and drier grass, juniper, and oak habitats (N = 272 and 72 captures, respectively) than they were in the riparian habitats (N = 79 and 11 captures, respectively). In contrast, the tree-foraging Clark spiny lizard (*Sceloporus clarkii*) and tree lizard (*Urosaurus ornatus*) were almost absent from the grass and juniper habitats (N = 10 and 2 captures, respectively) but were very abundant in the other areas (N = 150 and 122 captures, respectively). Similarly, of the 15 additional species caught in much lower numbers, several were obviously linked with specific habitats. For example, species found in specific habitats included Great Plains skinks (*Eumeces obsoletus*) in grass and juniper, lesser earless lizards (*Holbrookia maculata*) in juniper, madrean alligator lizards (*Gerrhonotus kingii*) in sycamore, mountain spiny lizard (*Sceloporus jarrovii*) in sycamore/juniper/oak, and black-necked garter snakes (*Thamnophis cyrtopsis*) in sycamore/juniper.

Thirteen species of small mammals were caught during the 2-year study. Desert shrews (*Notiosorex crawfordi*) were by far the most common small mammal species (N = 161, 65% of all captures). But 90% of these captures (N = 144) occurred in only 3 habitats—sycamore, sycamore/juniper/oak, and juniper. Eight species—northern pygmy mouse (*Baiomys taylori*), northern grasshopper mouse (*Onychomys leucogaster*), silky pocket mouse (*Perognathus flavus*), white-footed mouse (*Peromyscus leucopus*), deer mouse (*Peromyscus maniculatus*), western harvest mouse (*Reithrodontomys megalotis*), yellow-nosed cotton rat (*Sigmodon ochrognathus*), Arizona shrew (*Sorex arizonae*), and pocket gopher (*Thomomys* spp.)—were captured fewer than four times. Three species were relatively abundant (more than 10 captures), but were usually restricted to a few

TABLE 9.2.

Estimates for Amphibian, Reptile, and Small Mammal Populations in Riparian and Associated Habitats in Garden Canyon, Arizona, During Spring and Late Summer 1985 and 1986

	Habitat													
Species	Composite/ Grass		Sycamore		Juniper Woodland		Sycamore/ Juniper		Oak Woodland		Sycamore/ Juniper/Oak		Total	
	1985	1986	1985	1986	1985	1986	1985	1986	1985	1986	1985	1986	1985	1986
Amphibians and Reptiles														
Number of captures	40	68	37	35	84	51	83	41	130	96	72	40	446	331
Number of species	3	5	7	7	7	6	9	5	7	5	8	4	15	13
Small Mammals														
Number of captures	19	8	50	38	27	23	31	28	5	6	9	4	141	107
Number of species	5	4	8	6	6	4	4	4	2	3	3	2	10	9

habitats. Southern grasshopper mice (*Onychomys torridus*) were found only in grass and juniper habitats (N = 21), brush mice (*Peromyscus boylei*) were found only in oak and sycamore habitats (N = 11), and fulvous harvest mice (*Reithrodontomys fulvescens*) were primarily caught in the grass, juniper, and sycamore habitats (N = 23).

QUEEN CREEK

Most reptiles and amphibians (88.8%) were trapped in the desert (Table 9.3.). Nine species—greater earless lizard (*Cophosaurus texanus*), desert spiny lizard (*Sceloporus magister*), western banded gecko (*Coleonyx variegatus*), gila monster (*Heloderma suspectum*), regal horned lizard (*Phrynosoma solare*), black-necked garter snake (*Thamnophis cyrtopsis*), western blind snake (*Leptotyphlops humilis*), banded sand snake (*Chilomeniscus cinctus*), and night snake (*Hypsiglena torquata*)—were trapped only on the desert upland or wash. In contrast, the western ground snake (*Sonora semiannulata*) was trapped exclusively in the riparian zone, although this species was observed on the transects in the desert. Snake species observed on the transects in the desert, but not in the riparian, were tiger rattlesnake (*Crotalus tigris*), coachwhip (*Masticophis flagellum*), common kingsnake (*Lampropeltis getulus*), and gopher snake (*Pituophis melanoleucus*). The western diamondback rattlesnake (*C. atrox*) was the only snake species observed in both the desert and riparian zone. More than twice as many species (including snakes observed on transects) were captured or observed in the desert (17) as compared to the riparian (7) (Table 9.3.).

Bird densities were higher in the more structurally diverse habitats, such as the riparian interior and riparian edge, than in structurally simpler habitats of the desert wash and the desert upland (Table 9.3.). In fact, bird density on the desert upland was less than one-third of that found in the riparian edge and interior habitats (Table 9.3.). In contrast to the immediate drop in bird density from the riparian edge to the desert upland, density along the desert wash gradually decreased to that of the desert upland. Ten bird species—Cooper's hawk (*Accipiter cooperii*), willow flycatcher (*Empidonax traillii*), Bewick's wren (*Thryomanes bewickii*), ruby-crowned kinglet (*Regulus calendula*), solitary vireo (*Vireo solitarius*), warbling vireo (*Vireo gilvus*), yellow-rumped warbler, common yellowthroat (*Geothlypsis trichas*), summer tanager (*Piranga rubra*), and song sparrow (*Melospiza melodia*)—were found only in the riparian habitats. Eight additional species—yellow-breasted chat (*Icteria virens*), Bell's vireo (*Vireo bellii*), yellow warbler (*Dendroica petechia*), Wilson's warbler (*Wilsonia*

TABLE 9.3.
Estimates for Terrestrial Vertebrate Populations Behind Whitlow Dam, Queen Creek, Central Arizona, 1981 (Birds Only) and 1982

	Riparian		Desert	
Species	Interior	Edge	Wash	Upland
Amphibians and Reptiles				
Number of captures	15	23	179	125
Number of species	4	6	12	10
Birds				
1981				
Density (pairs/40 ha)	336	429	217	108
Number of species	22	25	30	24
1982				
Density (pairs/40 ha)	437	445	326	145
Number of species	22	25	30	24
Small Mammals				
Number of captures	2	12	49	19
Number of species	2	4	3	2

pusilla), ladder-backed woodpecker (*Picoides scalaris*), Lucy's warbler (*Vermivora luciae*), white-winged dove (*Zenaida asiatica*), and Abert's towhee (*Pipilo albertii*)—had their highest densities in the riparian island. In fact, bird populations in the riparian stand were at least three times those on the adjacent desert upland (for more detail, see Szaro and Jakle 1985).

Only six small mammal species—desert shrew (*Notiosorex crawfordi*), pallid bat (*Antrozous pallidus*), Arizona pocket mouse (*Perognathus amplus*), Bailey's pocket mouse (*Perognathus baileyi*), cactus mouse (*Peromyscus eremicus*), and white-throated woodrat (*Neotoma albigula*)—were caught during this study (Table 9.3.). Most mammals (58%) were trapped in the desert wash where Bailey's pocket mouse was the single most abundant mammal. Three species—the pallid bat, white-throated woodrat, and cactus mouse—were trapped only in riparian habitats; the Arizona pocket mouse was trapped only in desert habitats.

MANAGEMENT IMPLICATIONS

The ability of interacting species to maintain dynamically stable populations or even to persist varies with the differing abilities of species to compensate for effects of increased exploitation or competition (Cole 1987). Changes in community structure in response to environmental stress, whether natural or human related, can be useful in predicting the specific outcome of management regimes (Karr 1987), but they must be partitioned according to cause. However, because differences in response to grazing exclusion on the Rio de las Vacas undoubtedly were further confounded by critical climatic events, examination of the dynamics of this high-elevation montane riparian ecosystem over a single year could lead to some incorrect assumptions and conclusions. At the Rio de las Vacas, the probability for error would have been greatest if conclusions were based only on the study of grazing impacts on small mammals in 1986, when population densities were extremely low.

The varying patterns of vertebrate responses are not unique to high-elevation riparian systems in the arid Southwest. At the mid-elevation site in Garden Canyon, the relative importance of the six habitats was considerably different when looking only at small mammal or amphibian and reptile use. For example, oak woodland and sycamore/juniper/oak sites proved to be very important sites for tree-foraging reptiles. In contrast, only a few species of small mammals with low numbers were found on the same sites. Again, management decisions about these areas based on only one group or the other might lead to undesirable consequences.

Low-elevation areas, as along Queen Creek, apparently showed the same nonsynchronized patterns. At Queen Creek, 10 bird species were recorded only in this riparian island and 8 additional species had their highest densities in this recently created environment (Szaro and Jakle 1985). Using only response information for birds, researchers and land managers could make erroneous conclusions about the overall value of mitigating riparian losses solely by changes in vegetative structure. Despite dramatic increases in bird populations in response to increased structural diversity of the vegetation upstream from the dam (Szaro and Jakle 1985), amphibian, reptile, and small mammal populations did not respond in a similar manner (Szaro and Belfit 1986, 1987) (Table 9.3.). Populations and species richness of these three vertebrate groups were actually lower than in the surrounding desert habitats. Vertebrates from the surrounding desert habitat infrequently use the riparian island. Moreover, use of the site by desert species has been restricted by the riparian stand. Some riparian vertebrate species typical of areas similar in structural diversity to Queen Creek were not found. This may be due to biogeographic considerations or the limited mobility of these species. Data from Queen Creek indicate that restoration of disturbed riparian faunas might require reintroduction of species, in addition to changes in vegetative complexity, to replicate full vertebrate community structure and richness at this and other riparian areas (Szaro and Belfit 1986, 1987).

Results of these studies suggest that descriptive short-term case-history studies of only selected species, although valuable, lack generalization. However, they do point in part to the necessity of conducting research that has pretreatment or frame-of-reference information (Rinne 1988), encompasses watersheds (ecosystems), and is long term (>5 years) in duration. Treatments or manipulations (e.g., different grazing or timber-harvesting intensities) should be instituted, dictated largely by the pretreatment information on habitat and biota. Further study should look not only for cause-and-effect relationships, but also for the processes by which these occur (Szaro and Rinne 1988).

RECOMMENDATIONS AND CONCERNS

Systematic patterns in populations and communities may become apparent if studies are done on a sufficiently large scale or over a sufficiently long time (May 1984). Evaluating the effects of resource management on vertebrate communities depends on establishing the baseline structure and composition of these communities and their range of natural variability over time (Karr 1987). The potential for incorrect management decisions may be further compounded by basing them on studies that examine species and habitat relationships only for a single season or year. Because all biotic communities are dynamic, the central issue is to distinguish between changes in species composition that result from changes in the physical and biotic environment from those changes that result from individual species' responses to weather fluctuations (Karr 1987).

The goals of community ecology are threefold: (1) to determine the patterns of natural systems; (2) to explain them by discerning the causal processes; and (3) to synthesize these explanations as far as possible (Wiens 1984). Findings from community ecology, therefore, can contribute to the decision-making process for land managers. Often practical decisions are made in haste and in the absence of desired empirical information (May 1984). Ideally, incisive use of biological knowledge should allow predictions rather than only after-the-fact judgments (Karr 1987). Different models and recommendations can be made in the same (or, in some cases, different) geographical areas, so that the management regime assumes some aspects of a controlled experiment (May 1984). I, therefore, argue for the overall ecosystem approach to research and ultimately to manage riparian areas.

The basic requirements of scientific research are testable hypotheses, control of variables in time and space, replication of treatments, and an unbiased research area. The case-history study on the Rio de las Vacas typifies the difficulty of achieving these basic requirements even on a single stream on national forestlands. A researcher's lack of ability or authority to control the varying degree of multiple uses on national forestland precludes determining the differential impacts of these uses. Management of habitats and vertebrate species by different federal and state agencies on the same lands further complicates the situation.

To date, variations in results of research on grazing effects have resulted, in part, from faulty research designs (Platts and Nelson 1985) and, in part, from the arena available for study. All three study areas were recommended by forest or army biologists. Basic to effective land management are clearly defined goals and the use of sound information on the effects of management on habitats and biota. Land managers need answers to specific questions, but have been offered contradicting results of poorly designed studies. Both researchers and land managers must remedy the lack of reliable answers and the lack of suitable study areas. Absence of a mutual contract between these two parties will undoubtedly result in continued production of research results that are readily and validly criticized by individuals and user groups alike, and that most probably will not be applied in land-management activities (Szaro and Rinne 1988, Rinne 1989). In the end, it will be the resources, different in the eyes of each user, that will be affected adversely and, in some cases, exploited.

Acknowledgments

I thank D. R. Patton, J. N. Rinne, and S. C. Belfit for their constructive reviews. J. K. Aitkin, R. Babb, S. Belfit, H. Berna, M. Cady, C. Engel-Wilson, X. Hernandez, D. Johnson, M. Lane, W. Legarde, L. Simon, and D. Smith aided in the collection of the field data.

Literature Cited

Brady, W., D. R. Patton, and J. Paxson. 1985. The development of southwestern riparian gallery forests. Pages 39–43 *in* Johnson, R. R., C. D. Ziebel, D. R. Patton, P. F. Ffolliott, and R. H. Hamre, tech. coords. Riparian ecosystems and their management: reconciling conflicting uses. First North Amer. Riparian Conf. U.S.D.A. For. Serv. GTR-RM-120. Rocky Mtn. For. & Range Exp. Stn., Fort Collins, CO.

Burkham, D. E. 1972. Channel changes of the Gila River in Safford Valley, Arizona, 1846–1970. U.S. Geol. Surv. Prof. Pap. 655–G. 24pp.

Cannon, R. W., and F. L. Knopf. 1984. Species composition of a willow community relative to seasonal grazing histories in Colorado. Southwest. Nat. 29:234–237.

Cole, G. F. 1987. Changes in interacting species with disturbance. Environ. Manage. 11:257–264.

Fenner, P., W. W. Brady, and D. R. Patton. 1985. Effects of regulated water flows on regeneration of Fremont cottonwood. J. Range Manage. 38:135–138.

Heede, B. H. 1985. Interactions between streamside vegetation and stream dynamics. Pages 54–58 in Johnson, R. R., C. D. Ziebel, D. R. Patton, P. F. Ffolliott, and R. H. Hamre, tech. coords. Riparian ecosystems and their management: reconciling conflicting uses. First N. Amer. Riparian Conf. U.S.D.A. For. Serv. GTR-RM-120. Rocky Mtn. For. & Range Exp. Stn., Fort Collins, CO.

Hendrickson, D. A., and W. L. Minckley. 1984. Cienegas—vanishing climax communities of the American Southwest. Desert Plants 6:131–175.

Hupp, C. F. 1982. Stream-grade variation and riparian forest ecology along Passage Creek, Virginia. Bull. Tor. Bot. Club 109:488–499.

Hupp, C. R., and W. R. Osterkamp. 1985. Bottomland vegetation distribution along Passage Creek, Virginia, in relation to fluvial landforms. Ecology 66:670–681.

Johnson, R. R., and S. W. Carothers. 1982. Riparian habitats and recreation: interrelationships and impacts in the southwest and rocky mountain region. Eisenhower Consort. Bull. 12. U.S.D.A. For. Serv., Rocky Mtn. For. & Range Exp. Stn., Fort Collins, CO. 31pp.

Johnson, R. R., and D. A. Jones (tech. coords.) 1977. Importance, preservation, and management of riparian habitat: A symposium. U.S.D.A. For. Serv. Gen. Tech. Rep. RM-43. Rocky Mtn. For. & Range Exp. Stn., Fort Collins, CO. 217pp.

Jones, K. B. 1981. Effects of grazing on lizard abundance and diversity in western Arizona. Southwest. Nat. 26:107–115.

Karr, J. R. 1987. Biological monitoring and environmental assessment: a conceptual framework. Environ. Manage. 11:249–256.

Kauffman, J. B., and W. C. Krueger. 1984. Livestock impacts on riparian ecosystems and streamside management implications . . . a review. J. Range Manage. 37:430–438.

Knopf, F. L., and R. W. Cannon. 1982. Structural resilience of a Salix community to changes in grazing practices. Pages 298–307 in Peek, J. M. and P. D. Dalke, eds. Wildlife-livestock relationships symposium. Proceedings 10. Forestry, Wildlife, & Range Exp. Stn., Moscow, ID.

Kozlowski, T. T., editor. 1984. Flooding and Plant Growth. Academic Press, Orlando, FL. 356pp.

MacArthur, R. W., and J. W. MacArthur. 1961. On bird species diversity. Ecology 42:594–598.

Manning, R. E. 1979. Impacts of recreation on riparian soils and vegetation. Water Res. Bull. 15:30–43.

Martin, S. C. 1979. Evaluating the impacts of cattle grazing on riparian habitats in the national forests of Arizona and New Mexico. Pages 35–38 *in* O. B. Cope, ed. Proceedings of the Forum—Grazing and riparian/stream ecosystems. Trout Unlimited, Inc., Denver, CO.

May, R. M. 1984. An overview: real and apparent patterns in community structure. Pages 3–16 *in* D. R. Strong, Jr., D. Simberloff, L. G. Abele, and A. B. Thistle, eds. Ecological communities: conceptual issues and the evidence. Princeton Univ. Press, NJ.

McLeod, K. W., and J. K. McPherson. 1973. Factors limiting the distribution of *Salix nigra*. Bull. Tor. Bot. Club 100:102–110.

Minckley, W. L., and D. E. Brown. 1982. Wetlands. Pages 273–287 *in* D. E. Brown, ed. Biotic communities of the American Southwest—United States and Mexico. Desert Plants 4:1–342.

Minckley, W. L., and T. O. Clark. 1984. Formation and destruction of a Gila River mesquite bosque community. Desert Plants 6:23–30.

Minckley, W. L., and J. N. Rinne. 1985. Large woody debris in hot-desert streams: a historical review. Desert Plants 7:142–153.

Mosconi, S. L., and R. L. Hutto. 1982. The effect of grazing on the land birds of a western Montana riparian habitat. Pages 221–233 *in* Peek, J. M. and P. D. Dalke, eds. Wildlife-livestock relationships symposium: Proceeding 10. Forestry Wildlife & Range Exp. Stn., Moscow, ID.

Nilsson, C. 1982. Effect of stream regulation on riparian vegetation. Pages 93–106 *in* Lillehammer, A., and S. J. Saltveit, eds. Regulated rivers. Universitet Sforlaget As, Oslo.

Petts, G. E. 1984. Impounded rivers: perspective for ecological management. John Wiley & Sons, NY. 326pp.

Pickett, S. T. A. 1980. Non-equilibrium coexistence of plants. Bull. Tor. Bot. Club 107:238–248.

Platts, W. S., K. A. Gebhardt, and W. L. Jackson. 1985. The effects of large storm events on basin-range riparian stream habitats. Pages 30–34 *in* Johnson, R. R., C. D. Ziebel, D. R. Patton, P. F. Ffolliott, and R. H. Hamre, tech. coords. Riparian ecosystems and their management: reconciling conflicting uses. First N. Amer. Riparian Conf. U.S.D.A. For. Serv. GTR-RM-120. Rocky Mtn. For. & Range Exp. Stn., Fort Collins, CO.

Platts, W. S., and R.L. Nelson. 1985. Streamside and upland use by cattle. Rangelands 7:5–7.

Platts, W. S., and F. J. Wagstaff. 1984. Fencing to control livestock grazing on riparian habitats along streams: Is it a viable alternative? N. Amer. J. Fish. Manage. 4:266–272.

Rinne, J. N. 1988. Grazing effects on stream habitat and fishes: research design considerations. N. Amer. J. Fish. Manage. 8:240–247.

————. 1989. Minimizing livestock grazing effects on riparian stream habitats:

recommendations for research and management. Pages 1–13 *in* G. Flock, ed., Enhancing States' lake/wetland programs. N. Amer. Lake Manage. Soc., Chicago, IL.

Rotenberry, J. T. 1985. The role of habitat in avian community composition: physiognomy or floristics. Oecologia 67:213–217.

Szaro, R. C. 1980. Factors influencing bird populations in southwestern riparian forests. Pages 403–418 *in* Proceedings of the workshop on Management of Western Forests and Grasslands for Nongame Birds. U.S.D.A. For. Serv. Gen. Tech. Rep. INT–86. Intermountain For. & Range Exp. Stn., Ogden, UT.

Szaro, R. C. 1989. Riparian forest and scrubland community types of Arizona and New Mexico. Desert Plants 9:66–139.

Szaro, R. C., and S. C. Belfit. 1986. Herpetofaunal use of a desert riparian island and its adjacent scrub habitat. J. Wildl. Manage. 50:752–761.

Szaro, R. C., and S. C. Belfit. 1987. Small mammal use of a desert riparian island and its adjacent scrub habitat. U.S.D.A. For. Serv. Res. Note RM–473. Rocky Mtn. For. & Range Exp. Stn., Fort Collins, CO. 5pp.

Szaro, R. C., S. C. Belfit, J. K. Aitkin, and Randall D. Babb. 1988*a*. The use of timed fixed-area plot samples and a mark-recapture technique in assessing riparian garter snake populations. Pages 239–246 *in* R. C. Szaro, D. R. Patton, and K. E. Severson, eds. Management of small mammals, amphibians and reptiles in North America. U.S.D.A. For. Serv. GTR-RM-166. Rocky Mtn. For. & Range Exp. Stn., Fort Collins, CO.

Szaro, R. C., and L. F. DeBano. 1985. The effects of streamflow modification on the development of a riparian ecosystem. Pages 211–215 *in* R. R. Johnson, C. D. Ziebel, D. R. Patton, P. F. Ffolliott, and R. H. Hamre, tech. coords. Riparian ecosystems and their management: reconciling conflicting uses. First N. Amer. Riparian Conf. U.S.D.A. For. Serv. Gen. Tech. Rep. RM-120. Rocky Mt. For. & Range Exp. Stn., Fort Collins, CO.

Szaro, R. C., and M. D. Jakle. 1985. Avian use of a desert riparian island and its adjacent scrub habitat. Condor 87:511–519.

Szaro, R. C., and J. N. Rinne. 1988. Ecosystem approach to management of southwestern riparian communities. Trans. N. Amer. Wildl. & Nat. Res. Conf. 53:502–511.

Szaro, R. C., L. H. Simons, and S. C. Belfit. 1988*b*. Comparative effectiveness of pitfalls and livetraps in determining small mammal community structure. Pages 282–288 *in* R. C. Szaro, D. R. Patton, and K. E. Severson, eds. Management of small mammals, amphibians and reptiles in North America. U.S.D.A. For. Serv. GTR-RM-166. Rocky Mtn. For. & Range Exp. Stn., Fort Collins, CO.

Taylor, D. M. 1986. Effects of cattle grazing on passerine birds nesting in riparian habitat. J. Range Manage. 39:254–258.

Thomas, J. W., C. Maser, and J. E. Rodiek. 1979. Riparian zones in managed rangelands—their importance to wildlife. Pages 21–30 *in* O. B. Cope, ed. Proceedings of the Forum—Grazing and riparian/stream ecosystems. Trout Unlimited, Inc., Denver, CO.

Turner, R. M., and M. M. Karpiscak. 1980. Recent vegetation changes along the Colorado River between Glen Canyon Dam and Lake Mead, Arizona. Geol. Surv. Prof. Pap. 1132. 125pp.

Verner, J. 1986. Future trends in management of nongame wildlife: a researchers viewpoint. Pages 149–171 *in* J.B. Hale, L.B. Best, and R.L. Clawson, eds. Management of nongame wildlife in the midwest: a developing art. Proc. Symp. 47th Midwest Fish & Wildl. Conf.

Verner, J., and A. S. Boss, editors. 1980. California wildlife and their habitats: western Sierra Nevada. U.S.D.A. For. Serv. Gen. Tech. Rep. PSW–37. Pacific Southwest For. & Range Exp. Stn., Berkeley, CA. 439pp.

White, P. S. 1979. Pattern, process, and natural disturbance in vegetation. Bot. Rev. 45:229–299.

Wiens, J. A. 1984. On understanding a non-equilibrium world: myth and reality in community patterns and processes. Pages 439–457 *in* D. R. Strong, Jr., D. Simberloff, L. G. Abele, and A. B. Thistle, eds. Ecological communities: conceptual issues and the evidence. Princeton Univ. Press, NJ.

Wiens, J. A., and J. T. Rotenberry. 1981. Habitat associations and community structure of birds in shrubsteppe environments. Ecol. Monogr. 51:21–41.

List of Authors

Paul Alaback
Research Ecologist
Pacific Northwest Experiment Station
U.S.D.A. Forest Service

David A. Anderson
Regional Supervisor
Alaska Department of Fish and Game
Juneau, Alaska

John A. Bissonette
Professor of Fisheries and Wildlife Resources
    and Leader of the Utah Cooperative Fish
    and Wildlife Research Unit
College of Natural Resources
Department of Fisheries and Wildlife
Utah State University

Fred C. Bryant
Professor
Department of Range and Wildlife Management
Texas Tech University
Lubbock, Texas 79409-2125

Ted T. Cable
Associate Professor
Department of Forestry
Kansas State University

Jere Christner
Fish, Wildlife, and Watershed Staff Officer
Chatham Area
U.S.D.A. Forest Service

Wayne H. Davis
Professor
School of Biological Sciences
University of Kentucky

Richard M. DeGraaf
Principal Research Wildlife Biologist
Northeastern Forest Experiment Station
U.S.D.A. Forest Service
Amherst, Massachusetts 01003

Glenn D. DelGiudice
Department of Fisheries and Wildlife
University of Minnesota
St. Paul, Minnesota

Thomas DeMeo
Ecologist
Ketchikan Area
U.S.D.A. Forest Service

Arlene Dolye
Wildlife Habitat Relationships Coordinator
Northern Region
U.S.D.A. Forest Service

Rodney W. Flynn
Wildlife Biologist
Alaska Department of Fish and Game
Juneau, Alaska

Jerry L. Franklin
Bloedel Professor of Ecosystem Analysis
College of Forestry
University of Washington
Seattle, Washington

James D. Fraser
Professor
Department of Fisheries and Wildlife
Virginia Tech University

Richard J. Fredrickson
Wildlife Biologist
USNPS
Olympic National Park
Natural Science Studies
600 East Park Avenue
Port Angeles, WA 98362

J. Edward Gates, Ph.D.
Associate Professor of Wildlife Ecology
Appalachian Environmental Laboratory,
Center for Environmental and Estuarine Studies
The University of Maryland System
Frostburg, Maryland 21532

Jon Martin
Ecology Program Coordinator
Chatham Area
U.S.D.A. Forest Service

William C. McComb
Associate Professor
Department of Forest Science
Oregon State University

Kevin McGarigal
Research Assistant
Department of Forest Science
Oregon State University

James McKibben
Resources Staff Officer
Stikine Area
U.S.D.A. Forest Service

L. David Mech
Patuxent Wildlife Research Center
U.S. Fish and Wildlife Service
Laurel, Maryland

Mark Orme
Wildlife Biologist
Tongass Interdisciplinary Team
U.S.D.A. Forest Service

Fred B. Samson
Wildlife Program Manager
Alaska Region
U.S.D.A. Forest Service

John W. Schoen
Senior Biologist and Conservation Biology Coordinator
Alaska Department of Fish and Game
Juneau, Alaska

Ulysses S. Seal
Research Service
U. S. Veterans Affairs Medical Center
Minneapolis, Minnesota

Lana G. Shea
Habitat Biologist
Alaska Department of Fish and Game
Juneau, Alaska

Lowell Suring
Wildlife and Fish Habitat Relationships Coordinator
Alaska Region
U.S.D.A. Forest Service

Robert C. Szaro
Ecologist
U.S.D.A. Forest Service

Kenneth Thompson
Fish and Wildlife Staff Officer
Ketchikan Area
U.S.D.A. Forest Service

Brian J. Tucker
Wildlife Biologist
Newfoundland and Labrador Wildlife Division
Building 810, Pleasantville
P.O. Box 8700
St. John's, Newfoundland, Canada A1B 4J6

Bruce G. Wilson
District Ranger
Big Summet Ranger District
U.S.D.A. Forest Service

Index

About the Editors

JON E. RODIEK is a professor of landscape architecture and urban and regional planning in the College of Architecture at Texas A&M University. He has served as: Division Chair for Design Innovation; Associate Dean for Academic Affairs; and is currently Chair of the Dean's Council in the College of Architecture. He has degrees in Landscape Architecture (B.L.A., M.L.A.), Plant Science (B.S.), Forestry (M.S.), and Natural Resource Management (Ph.D.). His primary work has been in the field of land planning and design. For the last decade he has focused his attention on the protection and rehabilitation of wildlife and wildlife habitat.

ERIC G. BOLEN currently is Dean of the Graduate School at the University of North Carolina at Wilmington. He formerly was Paul Whitfield Horn Professor in the Department of Range and Wildlife Management at Texas Tech University. Dr. Bolen earned his B.S. at the University of Maine, and his M.S. and Ph.D. degrees at Utah State University. He is the author or coauthor of more than 150 publications, including a college-level textbook.

Also Available from Island Press

Ancient Forests of the Pacific Northwest
By Elliott A. Norse

Balancing on the Brink of Extinction: The Endangered Species Act and Lessons for the Future
Edited by Kathryn A. Kohm

Better Trout Habitat: A Guide to Stream Restoration and Management
By Christopher J. Hunter

The Challenge of Global Warming
Edited by Dean Edwin Abrahamson

Coastal Alert: Ecosystems, Energy, and Offshore Oil Drilling
By Dwight Holing

The Complete Guide to Environmental Careers
The CEIP Fund

Economics of Protected Areas
By John A. Dixon and Paul B. Sherman

Environmental Agenda for the Future
Edited by Robert Cahn

Environmental Disputes: Community Involvement in Conflict Resolution
By James E. Crowfoot and Julia M. Wondolleck

Fighting Toxics: A Manual for Protecting Your Family, Community, and Workplace
Edited by Gary Cohen and John O'Connor

Forests and Forestry in China: Changing Patterns of Resource Development
By S. D. Richardson

From *The Land*
Edited and compiled by Nancy P. Pittman

Hazardous Waste from Small Quantity Generators
By Seymour I. Schwartz and Wendy B. Pratt

Holistic Resource Management Workbook
By Alan Savory

In Praise of Nature
Edited and with essays by Stephanie Mills

218

The Living Ocean: Understanding and Protecting Marine Biodiversity
By Boyce Thorne-Miller and John Catena

Natural Resources for the 21st Century
Edited by R. Neil Sampson and Dwight Hair

The New York Environment Book
By Eric A. Goldstein and Mark A. Izeman

Overtapped Oasis: Reform or Revolution for Western Water
By Marc Reisner and Sarah Bates

Permaculture: A Practical Guide for a Sustainable Future
By Bill Mollison

Plastics: America's Packaging Dilemma
By Nancy A. Wolf and Ellen D. Feldman

The Poisoned Well: New Strategies for Groundwater Protection
Edited by Eric Jorgensen

Race to Save the Tropics: Ecology and Economics for a Sustainable Future
Edited by Robert Goodland

Recycling and Incineration: Evaluating the Choices
By Richard A. Denison and John Ruston

Reforming The Forest Service
By Randal O'Toole

The Rising Tide: Global Warming and World Sea Levels
By Lynne T. Edgerton

Rush to Burn: Solving America's Garbage Crisis?
From *Newsday*

Saving the Tropical Forests
By Judith Gradwohl and Russell Greenberg

War on Waste: Can America Win Its Battle With Garbage?
By Louis Blumberg and Robert Gottlieb

Western Water Made Simple
From *High Country News*

Wetland Creation and Restoration: The Status of the Science
Edited by Mary E. Kentula and Jon A. Kusler

For a complete catalog of Island Press publications, please write:
Island Press
Box 7
Covelo, CA 95428. Or call 1-800-828-1302